International Conference on Intelligent Manufacturing and Intelligent Materials (2IM 2019)

IOP Conference Series: Materials Science and Engineering
Volume 565

Sanya, China
9 – 11 May 2019

ISBN: 978-1-5108-9391-7
ISSN: 1757-8981

Printed from e-media with permission by:

Curran Associates, Inc.
57 Morehouse Lane
Red Hook, NY 12571

proceedings.com

Some format issues inherent in the e-media version may also appear in this print version.

Printed by Curran Associates, Inc. (2019)

Additional copies of this publication are available from:

Curran Associates, Inc.
57 Morehouse Lane
Red Hook, NY 12571 USA
Phone: 845-758-0400
Fax: 845-758-2633
Email: curran@proceedings.com
Web: www.proceedings.com

Table of contents

Volume 565

2019 International Conference on Intelligent Manufacturing and Intelligent Materials (2IM 2019)

9–11 May 2019, Sanya, China

Accepted papers received: 6 June 2019
Published online: 26 June 2019

Preface

Papers

IOP Conf. Series: Materials Science and Engineering **565** (2019) 011001 doi:10.1088/1757-899X/565/1/011001

PREFACE

2019 International Conference on Intelligent Manufacturing and Intelligent Materials (2IM 2019) was held in Sanya, China from May 9 to 11, 2019. 2IM 2019 was co-organized by is organized by Harbin Institute of Technology, Hong Kong Society of Mechanical Engineers (HKSME) and Australian-China Advanced Material and Manufacturing Association, sponsored by South China University of Technology. The conference provides a useful and wide platform both for display the latest research and for exchange of research results and thoughts in Intelligent Manufacturing and Intelligent Materials other topics. The participants of the conference were from almost every part of the world, with background of either academia or industry, even well-known enterprise. The success and prosperity of the conference is reflected high level of the papers received.

The proceedings are a compilation of the accepted papers and represent an interesting outcome of the conference. This book covers 2 chapters: Intelligent Manufacturing and Intelligent Materials.

We would like to acknowledge all of those who supported 2IM 2019. Each individual and institutional help were very important for the success of this conference. Especially we would like to thank the organizing committee for their valuable advices in the organization and helpful peer review of the papers.

We sincerely hope that 2IM 2019 will be a forum for excellent discussions that will put forward new ideas and promote collaborative researches. We are sure that the proceedings will serve as an important research source of references and the knowledge, which will lead to not only scientific and engineering progress but also other new products and processes.

Prof. Ruxu Du, South China University of Technology, China

IOP Conf. Series: Materials Science and Engineering **565** (2019) 011002 doi:10.1088/1757-899X/565/1/011002

COMMITTEES

Conference Co-Chair

Ruxu Du, South China University of Technology, China

Tao Sun, Harbin Institute of Technology, China

Program Chair

Junjie Zhang, Harbin Institute of Technology, China

Katz Zvi, University of Pretoria, South Africa

Local Committee

Yujun Li, Chengdu Sherlock Education Consultant Ltd., China

Suzy Shih, Hong Kong Society of Mechanical Engineers (HKSME), Hong Kong

International Technical Committee

Yongsheng Ma, University of Alberta, Canada

Ping Guo, The Chinese University of Hong Kong, HK

Raghu Prakash, Indian Institute of Technology Madras, India

Hamid Bahai, Brunel University London, UK

Ştefan Ţălu, The Technical University of Cluj-Napoca, Romania

Kean C Aw,The University of Auckland, New Zealand

Shoujin Sun, Swinburne University of Technology, Australia

Yuncang Li, RMIT University, Australia

Wong Yew Hoong, University of Malaya, Malaya

YANG Zhengbao, City University of Hong Kong

Khalid Arif, Massey University, New Zealand

Chow Yin Lai, RMIT University, Australia

Keng-Te Lin, National Taiwan University, Taiwan, ROC

Zuraidah Mohd Zain,Unversiti Malaysia Perlis, Malaysia

IOP Conf. Series: Materials Science and Engineering **565** (2019) 011003 doi:10.1088/1757-899X/565/1/011003

Peer review statement

All papers published in this volume of *IOP Conference Series: Materials Science and Engineering* have been peer reviewed through processes administered by the proceedings Editors. Reviews were conducted by expert referees to the professional and scientific standards expected of a proceedings journal published by IOP Publishing.

A Production Planning and Scheduling Method Based on Heuristic Rules for Forming-sintering Production System

Zhang Zhengmin, Guan Zailin, Yue Lei*, Wang Chuangjian and Wang Hao

School of Mechanical Science and Engineering, Huazhong University of Science and Technology, Wuhan, China

*Email: leileyok@hust.edu.cn

Abstract. The forming-sintering process is the bottleneck of the magnetic material manufacturing system in general, which restricts the overall output of the workshop. It is hard to formulate a production plan due to the complex constraints in such production system. In current research, the production planning control framework integrated with DBR (Drum-Buffer-Rope) concept oriented for forming-sintering process is designed. Moreover, a heuristic method considering the constraints of the magnetic material production process is proposed, which standardize and optimize the preparation process of the forming-sintering production plan. The practical application indicates that the heuristic method greatly reduces the number of WIPs, shortens the order manufacturing cycle, and improves the on-time delivery rate and the utilization rate of the sintering furnace, as compared with the current planning method of the enterprise.

1. Introduction

Due to the complexity of the magnetic material production process, some manufacturing processes are closely related and constrained, and the production planning of magnetic materials is extremely difficult to formulate [1]. At present, the production plan of magnetic material is mainly prepared by the planners based on experience, and each process's plan is compiled independently, resulting in a rough plan, which is difficult to implement to actual production system and guarantee the order delivery.

The forming-sintering process is the bottleneck of the magnetic material production system, which restricts the overall output of the workshop. Due to the complexity constraints of the two processes, it is hard to formulate a manufacture plan. Through the rational design of the forming-sintering production planning method, it is possible to standardize the preparation process of the forming-sintering production plan, improve the utilization of bottleneck resources, and reduce the number of WIPs. There were few studies related to the production planning or scheduling of magnetic materials, only a few research work from Northeastern University focused on the forming-sintering process separately[2,3]. For example, Liu [2] focused on process constraints and planning rules, but did not conduct in-depth study on the overall preparation method of the forming-sintering planning.

Due to the high complexity of the magnetic material production planning problem, the heuristic rule method is more suitable for finding a satisfactory solution to the production scheduling problem. The rule was based on a summary of past experience and can provide a satisfactory solution to the problem in a shorter period of time. Literature [4-8] provided a reference for the selection of heuristic rules and inspirations for the combination of heuristic rules and other mathematical methods.

2. Current status of magnetic material production

The manufacture of magnetic materials is a typical multi-variety and small batch production mode, includes two major processes: powder preparation and magnetic core production[9]. There are four processes in magnetic core production, which are forming, sintering. grinding, separating[10].The forming and sintering processes are two key processes for the production of magnetic core. The forming process is the first process of magnetic core production. The process requires a forming machine to shape the powder. The formulation of the forming production plan needs to consider the equipment capacity, constraints of the forming process, the powder inventory, as well as the roughcast arrangement constraints of the sintering process. Sintering is the bottleneck process for magnetic core production, and its equipment capacity limits the overall output of the plant. When formulating the sintering production plan, the main consideration is the arrangement of roughcast. The resources used in the sintering process are sintering furnaces, different heights in the furnace have different temperatures, different product types need to be placed at different layers in the furnace according to manufacturing requirements [11]. Therefore, it is necessary to arrange the products in groups before processing, that is, each layer is required to place with different products and fire into the sintering furnace simultaneously, the loading tool used for the arranged roughcasts is the pusher. At present, the production schedules of the two processes are formulated separately in actual production, and the products of the forming process cannot meet the requirements for the roughcast arrangement, resulting in too much buffering of the semi-finished products before the sintering process and lowly utilization rate of the sintering furnace .The utilization rate of the sintering furnace is an important index for the enterprise to measure whether the sintering blank is optimal. Therefore, the comprehensive optimization of the forming-sintering process is a significant way to improve the magnetic material production efficiency.

3. Control framework of forming-sintering production planning

There are many constraints between forming and sintering process. The production plan should consider a series of constraints, such as order delivery, customer priority, powder stock, equipment capacity, operation group capacity, product-mold correspondence, and roughcast arrangement rules. Planning solutions should determine the resource assignment for each process of each order, and the start/end time of each order. The strong constrained process characteristics require that the scheduling of these two processes should be considered together. Therefore, based on the DBR model[12] , a control framework of forming-sintering production plan has been proposed. Figure 1 shows the control frame diagram of forming-sintering. Firstly, according to the order requirements and the roughcast arrangement constraints, the order grouping plan can be formulated, then, the roughcast arrangement plan can be pulled by the order grouping plan, finally, the forming production plan can be prepared according to the roughcast arrangement planning and the constraints in the forming production process. According to the above processes, the maximum output of the bottleneck can be guaranteed.

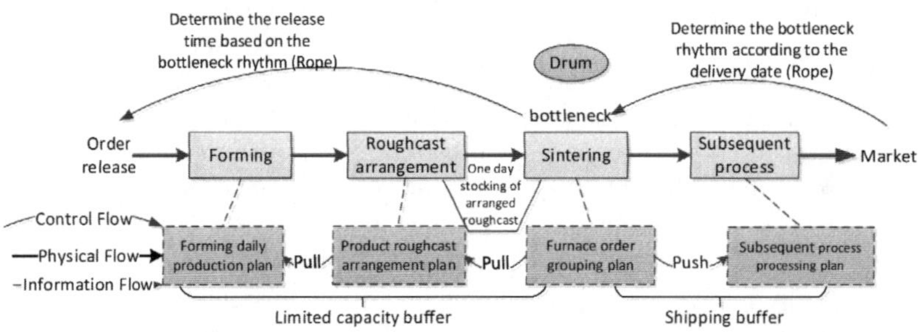

Figure 1. The control framework of forming-sintering production plan.

The production status of the forming-sintering process includes various constraints such as product, production capacity, inventory, and process. It is more complicated than the traditional mixed flow

IOP Conf. Series: Materials Science and Engineering **565** (2019) 012001 doi:10.1088/1757-899X/565/1/012001

production system. The traditional production models and scheduling algorithms are not suitable for solving this problem. After actual research and summary, this paper combines the production status of magnetic materials to establish a production model for magnetic material forming-sintering process, which considering delivery, powder inventory, equipment capacity, operation group capacity, process constraints and other constraints, and proposes a heuristic-based production planning method that can solve the production model.

4. Modeling of forming-sintering problem

4.1. Key elements

There are three key elements involved in this forming-sintering production planning problem:

(1) Decision variables: the production quantity of the production order in a certain shift; The number of furnaces of a certain shift.

(2) Target: minimize the production manufacturing cycle, produce higher priority products firstly .

(3) Constraints

1) Order delivery constraint. The forming-sintering production schedule needs to meet the due date.

2) Powder stock constraint. The daily used powder cannot exceed the current stock.

3) Roughcast arrangement constraint. Different types of orders should be placed at different heights of the sintering furnace and fired at the same time.

4) Mold capacity constraint. There is only one pair of molds on the production site, and the same mold only can be processed on one machine in a shift. When there are multiple orders to select the same mold for processing, the maximum capacity that can be processed in a shift is the production capacity of one forming machine.

5) Forming machine selection constraint. Different forming machines can be equipped with different specifications of the mold. Firstly, the selected mold is determined by the product type, and forming machine is processed and produced according to the selected mold. Different types of forming machines have different production capacities and need to be selected reasonably according to the size of the order.

6) Operation group capacity constraint. The forming equipment is responsible for the workers in the operation group, and each worker can only operate one forming machine, so the maximum forming capacity per day is limited by the capacity of the operation group.

4.2. Planning model

Table 1 is the relevant parameters and descriptions required by the model.

Table 1. List of parameter descriptions.

Subscript			
i,j	Order number, i=1,2…I	p	Type of molding machine, $p = 1,2 \ldots P$
a	The type of mold, , a=1,2…A	k	Powder type, k=1,2…K
t	Production days, t=1,2…T		
Set			
A_i	The set of molds that can be used by order i	P_a	Collection of molding machine P that can be used in mold a
I_k	Collection of orders that need to use powder k		
Parameter			
h	The length of time required for sintering a pusher	I	Number of order types
P	Number of molding machine types	A	Number of mold types
K	Number of powder types	T	Planning cycle

O_i	Production work order i	d_i	Due date of order i
C_P	Daily production capacity of molding machine p	C_{max}	Maximum daily capacity of the operation group
C_{min}	Minimum daily capacity of the operation group	L_i	The priority of the order i
M_i	The total modulus required for the order i	M_{is}	The remaining modulus need to be produced in order i
B_i	The arranging modulus of a pusher when the order i is sintered	W_i	The weight of the powder required for each model
F_{tk}	Stocks of type k powder on day t		
Variable			
X_{ik}	Whether the power's type that order i use is k,X_{ik}=1 means yes	X_{it}	If order i is processed on day t, thenX_{it}=1,Otherwise X_{it}=0
a_i	The type of mold used in order i	p_i	Type of forming machine used in order i

4.2.1. Decision variables. M_{it}: Indicates the modulus of the order i produced on day t. Y_t: Number of sintering furnaces on the day t.

4.2.2. Model expression. Object function:

$$\min \textstyle\sum_{i=1}^{I} Z_i L_i \tag{1}$$

Where:

$$Z_i = \begin{cases} 1 & \sum_{t=1}^{d_i} M_{it} < M_i \\ 0 & \sum_{t=1}^{d_i} M_{it} \geq M_i \end{cases} \tag{2}$$

s.t.

$$\textstyle\sum_{t=1}^{T} M_{it} \geq M_i \tag{3}$$

$$\textstyle\sum_{i=1}^{I} X_{ik} W_i M_{it} \leq F_{tk} \tag{4}$$

$$\textstyle\sum_{i=1}^{I} X_{ip} M_{it} \leq C_p \tag{5}$$

$$X_{ip} = \begin{cases} 1 & p_i = p \\ 0 & p_i \neq p \end{cases} \tag{6}$$

$$a_i \in I_a \text{且} p_i \in A_p \tag{7}$$

$$C_{min} \leq \textstyle\sum_{i=1}^{I} M_{it} \leq C_{max} \tag{8}$$

The second term of objective function (1) represents: Z_i in formula (2) is a parameter that reflects whether the production of the product meets the due date. If product i can be completed on time, then $Z_i = 1$, otherwise, $Z_i = 0$. The first term of objective function (1) represents: minimize total tardiness in the planning cycle, orders with higher priority need to be completed first. Equation (3) represents: all orders must be completed within period T, $i = 1,2 \dots I$. Equation (4) represents: the amount of each powder used per day cannot exceed the stock of the powder on that day.$k = 1,2,3,4; t = 1,2 \dots T$. Equation (5) represents molding capacity constraints. Equation (6) represents: the output of all orders using the forming machine p on day t cannot exceed the maximum capacity of the forming machine on that day, $p = 1,2 \dots P; t = 1,2 \dots T$.Equation (7) represents forming machine equipment selection constraint. Equation (8) represents the maximum forming capacity per day is affected by the operation group.

5. Algorithm implementation

Due to the complexity of the constraints, the intelligent algorithm is not suitable for this problem. The production planning algorithm based on heuristic rules can get a better solution at a faster rate and has a strong practicability.

5.1. Detailed rules

Based on a detailed analysis of the forming-sintering process, nine types of heuristic rules were established.

(1) Order sorting rules (R1)

1) According to the order delivery constraints, orders with the closest delivery date are preferred, in order to minimize the order tardiness.

2) If two orders have the same delivery date, the production orders should be arranged according to the order priority, in order to ensure the priority service of the important customers.

3) If two orders have the same delivery date and priority, the order with less demand modulus should be arranged first to shorten the production cycle. The specific rules are as follows:

R1.1: IF $d_i > d_j$, THEN prioritizing O_j

R1.2: IF $d_i = d_j, C_i > C_j$, THEN prioritizing O_i

R1.3: IF $d_i = d_j, C_i = C_j, M_i > M_j$, THEN prioritizing O_j

(2) Roughcast arrangement and combination rules (R2)

Considering the roughcast arrangement requirements in sintering, the roughcasts arranged in the same batch need to include products that are processed in all layers of the sintering furnace. The layer of the sintering furnace is represented by n, and the order to be placed in layer n is formed into a set. The process department will give the roughcast arrangement information table IT1 according to the order arranged by the rule R1.Then, according to the table IT1, the combined order is successively arranged for production. The formula expression is as follows.

R2: IF $OST_{R1} = \{O_1, O_2, O_3, \cdots, O_9, O_{10}, \cdots\}, IT1 = \begin{cases} O_3, O_4, O_7, O_{10} \\ O_1, O_2, O_6, O_9 \\ \cdots \end{cases}$, THEN $OST_{R21} = \{O_1, O_2, O_6, O_9, O_3, O_4, O_7, O_{10} \cdots\}$

(3) Roughcast grouping rules (R3)

When making the sintering plan, according to the order information table and the roughcast arrangement information table IT1, select the current unordered combined order at the front of the order list to be fired in a sintering furnace. The formula expression is as follows.

R3: IF $OST_{R21} = \{O_1, O_2, O_6, O_9, O_3, O_4, O_7, O_{10} \cdots\}, IT1 = \begin{cases} O_3, O_4, O_7, O_{10} \\ O_1, O_2, O_6, O_9 \\ \cdots \end{cases}$, THEN $Oven_1 = \{O_1, O_2, O_6, O_9\}, Oven_2 = \{O_3, O_4, O_7, O_{10}\}, \cdots$

(4)Sintering capacity calculation rule (R4)

There are F sintering furnaces in the sintering workshop, and each of them can fire all the magnetic cores. Each sintering furnace processes r pushers on average every day. After determining the roughcast arrangement method on the sintering furnace, the production modulus of a product can be multiplied by the number of pushers that can be sintered in the sintering furnace one day to calculate the production modulus of the product in one day. At the same time, the cycle that the order needs to be processed in the sintering process can be obtained. The specific rules are as follows:

R4.1: On day t,$B_i = x, (r \times x) \leq M_{iS}$, THEN $M_{it} = r \times x$,ELSE $M_{it} = M_{iS}$

R4.2: IF on day t,$B_i = x, (r \times x) > M_{iS}$, THEN after the order is completed, the next order for sintering the layer is allocated.

Note: Ceil (x) is an integer function.

(5) Powder inventory inspection rules (R5)

The total weight required for a type of powder on a given day should less than the stock of that type of powder on that day. In this planning model, when the sintering production plan is scheduled, the number of moulded products per day has been determined. Therefore, the inventory of powder should be checked when making the sintering plan, in order to prevent the shortage of powder stock. The specific rules are as follows:

R5: IF I_{tk} indicate the order set that required to use the powder type k on day t, THEN $\sum(M_{it} \times w_i) \leq F_{tk}, i \in I_{tk}$

(6) Production order splitting rules (R6)

After completing the roughcast arrangement plan of sintering. Determine the production sequence of sintering furnace and divide the orders into several transfer orders according to the production capacity of sintering furnace. Each transfer work order is the production work order of the day before the forming process, to achieve the goal of pulling the forming output completely by sintering demand. The specific rules are as follows:

R6: $O_{is} = T_{is} = \mathrm{Ceil}\left(\frac{M_i}{M_{it}}\right)$

(7) Mold selection rules (R7)

Each order type should choose the corresponding mold. The specific rules are as follows:

R7:$a_i \in A_i$

(8) Forming machine selection rule (R8)

R8:$p_i \in P_{ai}$

(9) Forming machine capacity constraint rule (R9)

The maximum daily production capacity of the forming machine varies according to the choice of the forming machine. The actual production capacity can not exceed the maximum capacity of the forming machine. The specific rules are as follows:

R9:$\sum_{i=1}^{I} X_{ip} M_{it} \leq C_p, X_{ip} = \begin{cases} 1 & p_i = p \\ 0 & p_i \neq p \end{cases}, p = 1,2 \ldots P; t = 1,2 \ldots T$

5.2. Forming-sintering production planning algorithm based on heuristic rules

The heuristic algorithm in this paper is based on the DBR strategy, which drives the whole production plan according to bottleneck production plan to ensure the maximum utilization of the bottleneck. According to the actual situation of enterprises, the first day's output of forming was used as a raw material for the second day of roughcast arrangement, that is, there was a one-day buffer between forming and sintering. In the planning period T, the sintering plan need to be prepared according to the roughcast arrangement information table IT1 and the order sorting information table IT2.Firstly, after carrying out the forming capacity evaluation and the powder inventory evaluation, the sintering production work order information table is generated in units of days ,and be used as input work orders for the forming schedule. The forming production plan should meet the demand of the sintering production work orders finally. The forming-sintering production planning steps are shown in Figure 2, the details steps are as follows:

(1) Order sorting stage

Step1: Sort all the orders that need to be processed by forming-sintering according to R1.1,get the sorted order sequence.

Step2: Sort the order sequence that obtained by Step 1 according to R1.2

Step3: Sort the order sequence that obtained by Step 2 according to R1.3

Step4: Obtain the roughcast arrangement information table IT1,sort the order sequence that obtained by Step3 according to R2.

(2) Production planning and scheduling stage

On the premise that the production order is sorted, the production-planning method for forming-sintering process based on heuristic rules has several steps:

1) Sintering planning stage:

Step1: Obtain the order sorting information table IT2 that was sorted already; t=2, indicate that the second day of the planning cycle T is the start date of the sintering roughcast arrangement plan

Step2: F=f, indicate that the available sintering furnace resources are f on day t.

Step3: According to table IT1 and IT2,obtain the orders that need to be processed by the current furnaces by R3.

Step4: Obtain the modulus of product i that can be sintered per pusher according to product's size and arrangement, then determine the modulus of the product i be sintered one day by rule R4.1

Step5: Correct the processed product results obtained by R3 according to R4.2

Step6: Update sintering blanking schedule table, obtain product categories and production information in current plan day, and divide each planned order into multiple transfer orders by sintering's daily production

Step7: F=F-1. Judge if F is 0, If yes, execute sequentially, otherwise get back to Step3

Step8: Get powder stock information table, determine whether the powder stock meets the demand according to rule R5; If yes, execute sequentially, otherwise back to Step3 and change order sequence.

Step9: t=t+1; If t≥T, execute sequentially. If t<T, get back to Step2.

Step10: Formulat a sintering roughcast arrangement plan, obtain the sintering roughcast arrangement plan table IT3; generate the transport order by rule R6,and use the daily total transfer order as the production work order the day before the forming process plan.

2) Forming planning stage:

Step11: t=1, indicate the first day of a planning cycle T is taken as the start date of the forming plan

Step12: i=1, indicate the first work order

Step13: Obtain the forming process's daily production information table (IT4), and select a feasible mold according to the rule R7;

Step14: Updat the forming process's production information table (IT4), which include the information of mold type, production modulus and so on.

Step15: According to the rule R8, select the type of forming machine.

Step16: Obtain the forming work order schedule

Step17: Judge whether the daily production capacity meets the production demand according to the rule R9: If yes, execute sequentially, otherwise, get back to Step13;

Step18: i=i+1.Judge if i is larger than I, If yes, execute sequentially, otherwise, get back to Step13;

Step19: Update the forming-sintering process scheduling results;

Step20: t=t+1.Judge if t is less than T, If yes, execute sequentially, otherwise, get back to Step11;

Step21: End.

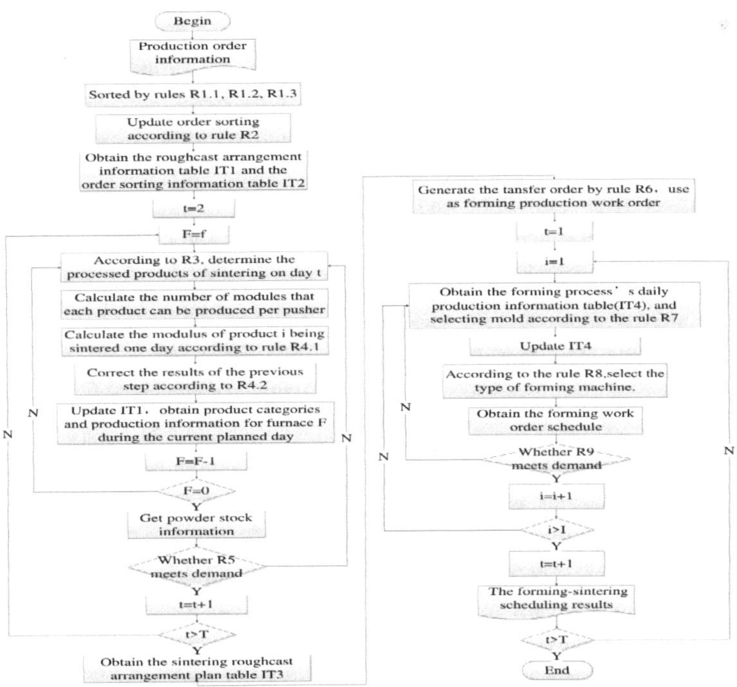

Figure 2. Flow chart of forming-sintering production planning.

6. Conclusion

In current research, a planning and scheduling problem in forming-sintering production environment is studied, and a heuristic method consisting of several optimal rules is proposed for the special planning problem. The following conclusions have been obtained after applying this method to a magnetic material production enterprise: (1) The method has a strong practicality and can be used to solve actual production problems in a short scheduling period. (2) It has a strong versatility. By summarizing the production characteristics and practical constraints of kinds of multi-process magnetic materials production planning problems, a comprehensive set of heuristic rules are established, which can be applied to most magnetic material production workshops.(3)The optimization performance is significant. It can be found that the overall output of the workshop has been greatly improved, the number of WIPs have been significantly reduced, the order manufacturing cycle has been shortened, and the production efficiency of the workshop has been greatly improved.

The heuristic concept can also be extended to the furnace grouping optimization problem in the entire material industry [13]. It is demonstrated that this method can make positive effect for the planning and scheduling decisions, i.e., shortening the order manufacturing cycle, reducing the number of WIPs, and improving the on-time delivery capability of orders.

References

[1] Zhang J, Wang Y, He H. Development prospect of soft ferrite materials in China[J]. *Journal of Materials and Power Sciences*, 2012(6):101-113.

[2] Liu Y, Pan Q, Chai T. Design and development of optimization software of magnetic multi-process coordination production planning based on benefit[C].*Intelligent Control and Automation*. IEEE, 2015:3428-3433.

[3] Liu Y, Xu G, Pan Q. Magnetic material molding sintering production scheduling optimization method and its application [C].*Journal of Zhejiang University(Engineering Science)*. 2013,47(9): 1517-1523

[4] Natarjan K, Mohanasundaram K M,et al. Performance evaluation of priority dispatching rules in multi-level assembly job shops with jobs having weights for flow-time and tardiness. *International Journal of Advances Manufacturing Technology*, 2007(31):p.751-761.

[5] Holthaus O, Rajendran C. Efficient dispatching rules for scheduling in a job shop.*International Joural of Production Economics*, 1997(48):p.87-105.

[6] Canbolat Y B, Gundogar E. Fuzzy priority rule for job shop scheduling. *Journal of Intelligent Manufacturing*, 2004(15):p.527-533.

[7] Huang Z. Workshop operation olanning algorithm and simulation based on heuristic scheduling rules [D]. *Wuhan: Huazhong University of Science and Technology*, 2009.

[8] Jiao Z, Xing Y. Clamping-sequence optimisation based on heuristic algorithm for sheet-metal components[J]. *International Journal of Production Research*, 2017(1):1-11.

[9] Paltanea G, Manescu Paltanea V, Gavrila H, et al. Magnetic Materials Used in the Magnetic Core Manufacture of Electrical Machines and Transformers[J]. *Key Engineering Materials, 2018, 775.*

[10] Xu G.Research and Development onMolding-Sintering Production PlanningSoftware of Sintering NdFeB Based on HeuristicRules [D]. *Northeast: Northeastern University*, 2012.

[11] Wang P S, Hou Y H, Hou J Y, et al. Development of Experimental Sintering Furnace for Mn-Zn Soft Ferrite Core[J]. *Mining & Metallurgical Engineering*, 2008.

[12] Michelsen A U. Theory of constraints (TOC)[M]. *Springer US*, 2000.

[13] Liu Y, Pan Q, Chai T. Method and application of production planning for sintered NdFeB belt[J]. *Control Engineering*, 2013, 20(4): 589-592.

Collection and Storage of Assembly Quality Data of Complex Electromechanical Products

Y Liang[1, 2*], Q D Zhang[1], N Zhao[1] and H Duan[2]

[1] School of Mechanical Engineering, University of Science and Technology, Beijing, 100083, China
[2] Beijing Aerospace Xinfeng Machinery Equipment Co., Ltd., Beijing, 100083, China

[*]E-mail: 511513881@qq.com

Abstract. In the context of made in China 2025 and intelligent manufacturing, "data-based decision" is one of the important concepts for quality management promotion, and data integrity and availability is an important aspect of data collection and data management. Combined with the characteristics of complex electromechanical products with complex structure, multiple development links and high quality requirements, this paper discusses the management category of quality big data from the perspective of manufacturing process operation management and product life cycle, studies the acquisition strategy by manual and automatic methods, acquisition process, the quality data modeling and expression specifications based on unified coding specifications and business flow and data flow framework and provides references for enterprises to carry out quality big data construction.

1. Introduction

In the context of made in China 2025 and intelligent manufacturing, digitization has become the basic work to which every enterprise pays more and more attention. In the fields of quality management, six sigma and continuous improvement, the "data-based decision" is one of the important concepts to promote [1]. Because of that, at present, a great number of studies have been carried out in the lean quality management in various fields and obtained certain achievements. On the basis of previous studies, combined with the complex structure, many development links and high quality requirements of the complex electromechanical products, this paper carries out the study on the collection and storage of multi-source heterogeneous quality data to provide data support for the enterprises to improve product quality, process quality and work quality.

The architecture of the main business of the manufacturing enterprise related to manufacturing is shown in Figure 1. Among them, production scheduling, production control, quality assurance, material and energy control, product inventory management and equipment maintenance management are the main businesses of manufacturing process operation management, while order management, cost accounting, procurement, R&D and marketing and other non-main businesses have an inseparable relationship with the manufacturing process [2].

IOP Conf. Series: Materials Science and Engineering **565** (2019) 012002 doi:10.1088/1757-899X/565/1/012002

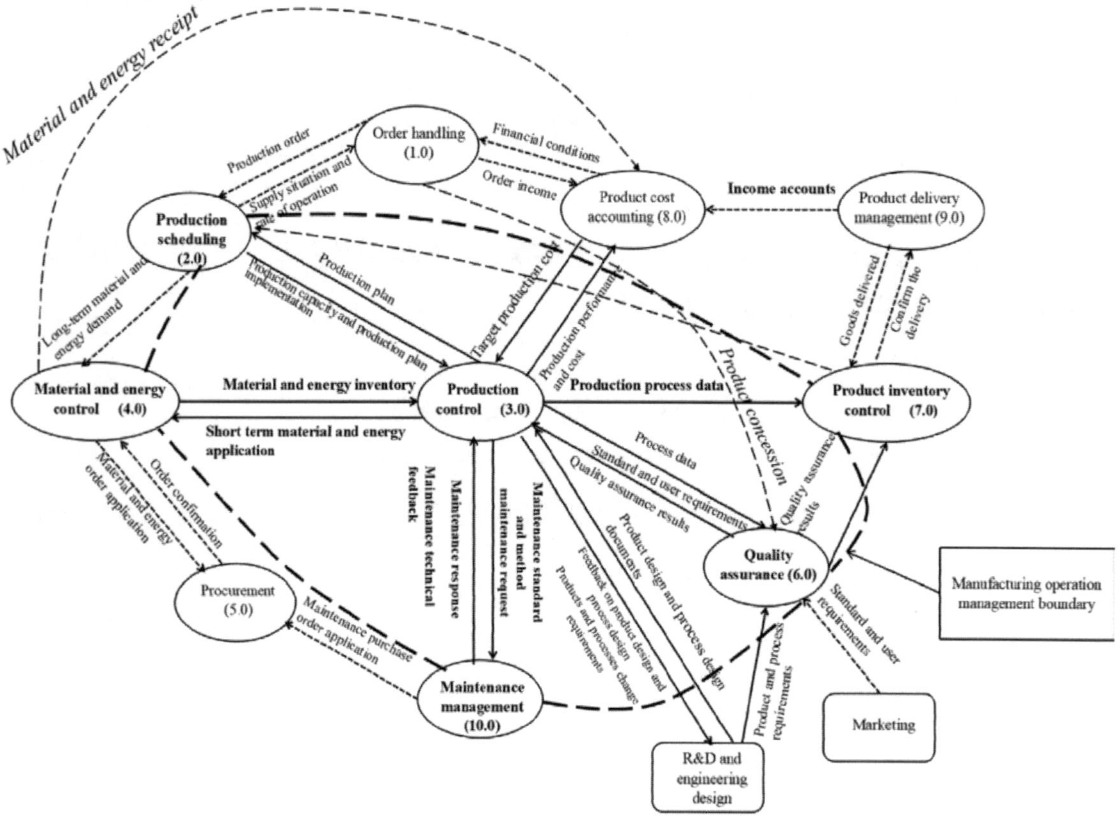

Figure 1. Manufacturing enterprise control function model

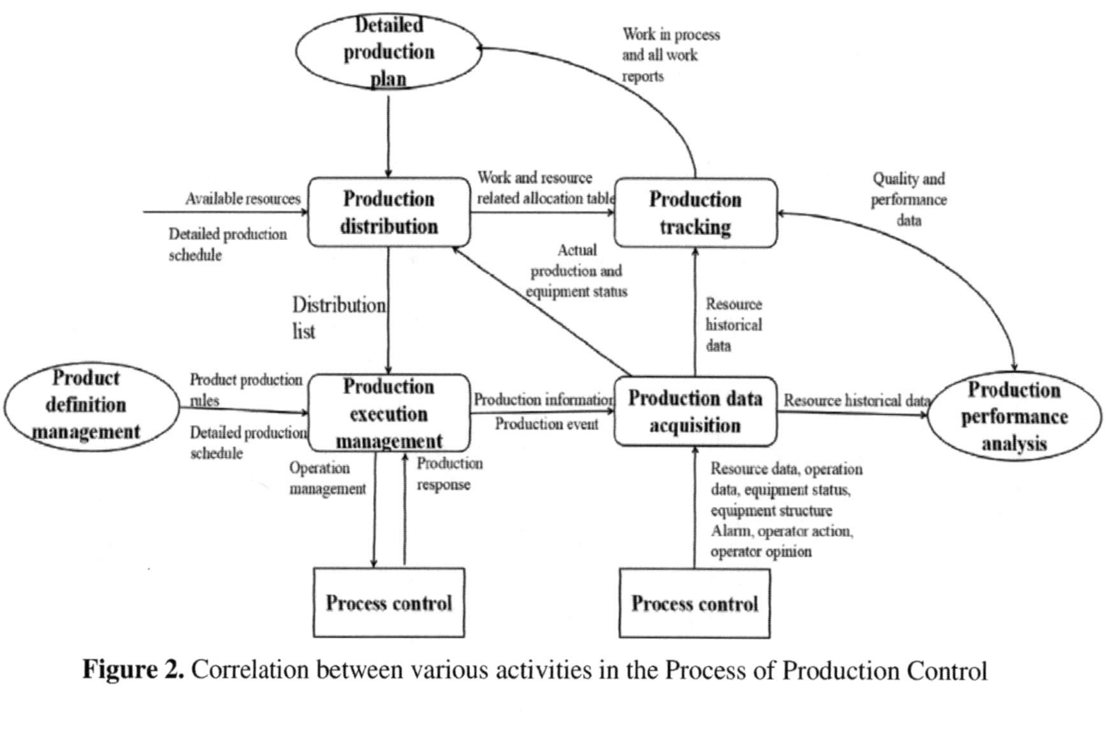

Figure 2. Correlation between various activities in the Process of Production Control

The activity of "production control" is to control the production, processing, assembly, debugging and other activities according to the production plan, that is, to monitor the man, machine, material, method, measurement and environment (5M1E) [3]. The fluctuation of 5M1E in the production process centrally reflects the implementation of the production plan [4]. On the one hand, the resource data, operation data, equipment status, quality problems and other information collected through production data collection activities can be used for production performance analysis. On the other hand, the documentary of production process can be used for production process tracking and is the basis of process capability evaluation and adjustment. The correlation between the various activities in the process of production control is shown in Figure 2.

2. Management category of quality big data

In the narrow sense, the quality data refers to the data related to product quality, such as the measured value of key characteristics, percent defective, percent of pass and first pass yield [5]. In the broad sense, the quality data reflects the product quality, process quality and work quality, such as fluctuation of process parameters, equipment operation condition, quality cost loss, production batch, inventory, invalid operation time, production plan completion, process change, equipment downtime loss, supply quality fluctuation and material shortage time, which will become the study objects of enterprise lean quality management. Therefore, the management category of quality big data refers to the basic information, status information, result information and analysis and evaluation information covering various business activities in each stage of product formation. From the perspective of manufacturing process operation management, the information included in quality big data is shown in Figure 3:

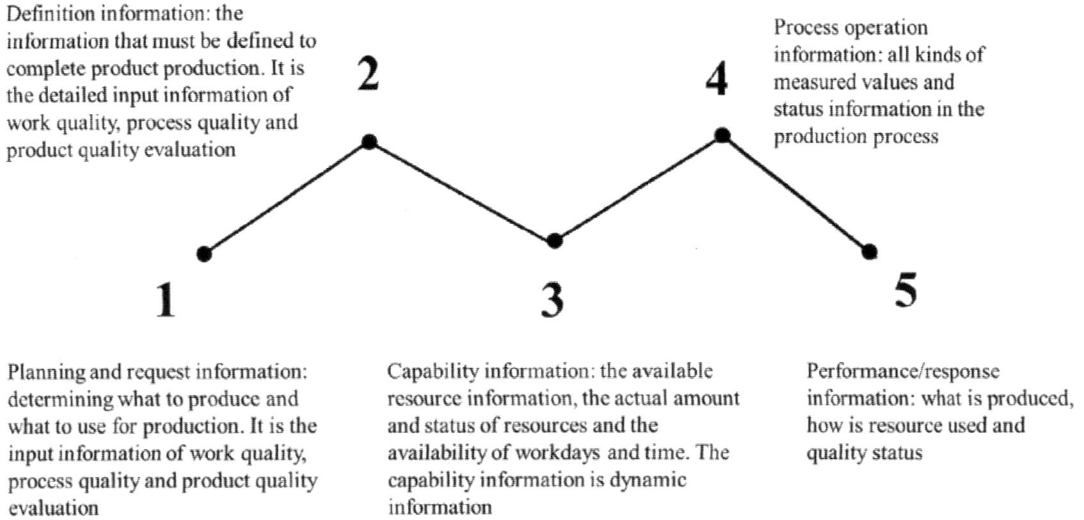

Figure 3. Information included in quality big data

The information in Figure 3 is both hierarchical and interrelated, and the correlation between the levels is shown in Figure 4. From the order placing, production plan preparation and plan execution to the collaborative operation of the 5M1E and other resources and production factors in the processes as well as the operation process and results, the organic whole shall be driven by the top-down production orders and plans on the one hand and achieves the short-term, medium-term and long-term analysis and evaluation of the whole manufacturing operation by perceiving the product, process, equipment and tool status and collecting the operating data on the other hand, so as to realize the closed loop control of the production process.

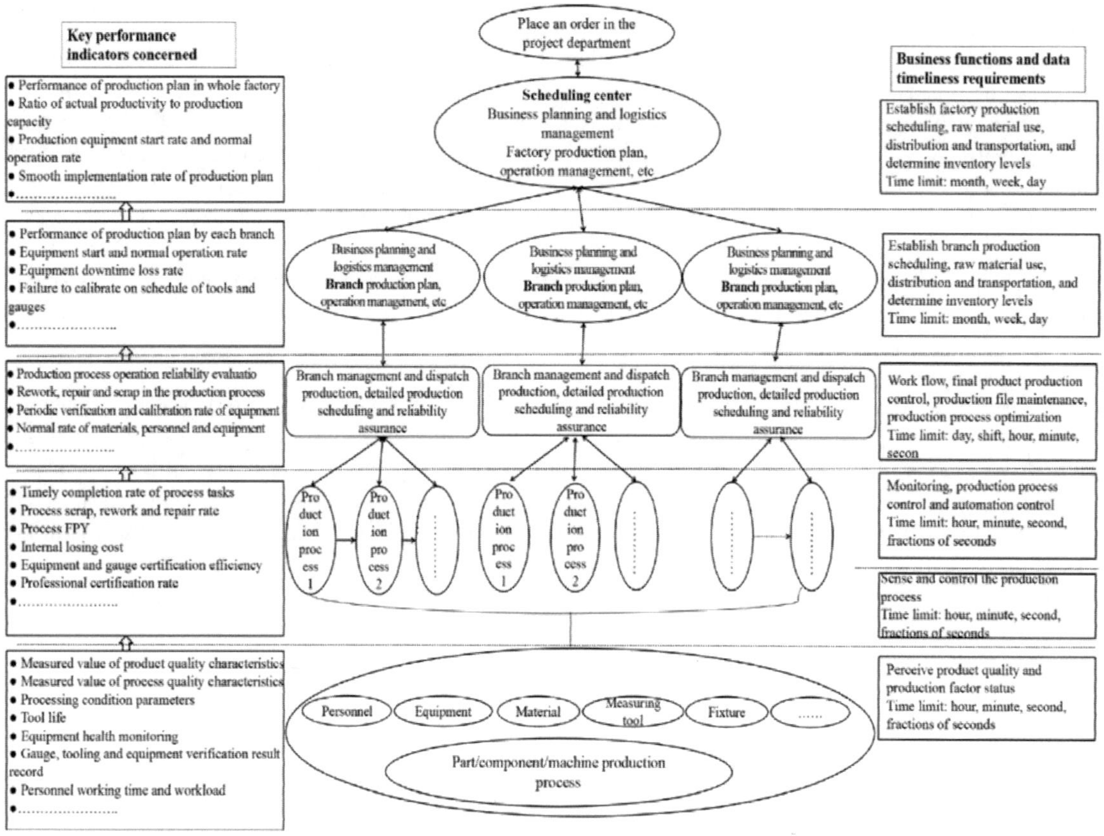

Figure 4. Hierarchical relationship between quality big data and manufacturing operation management process

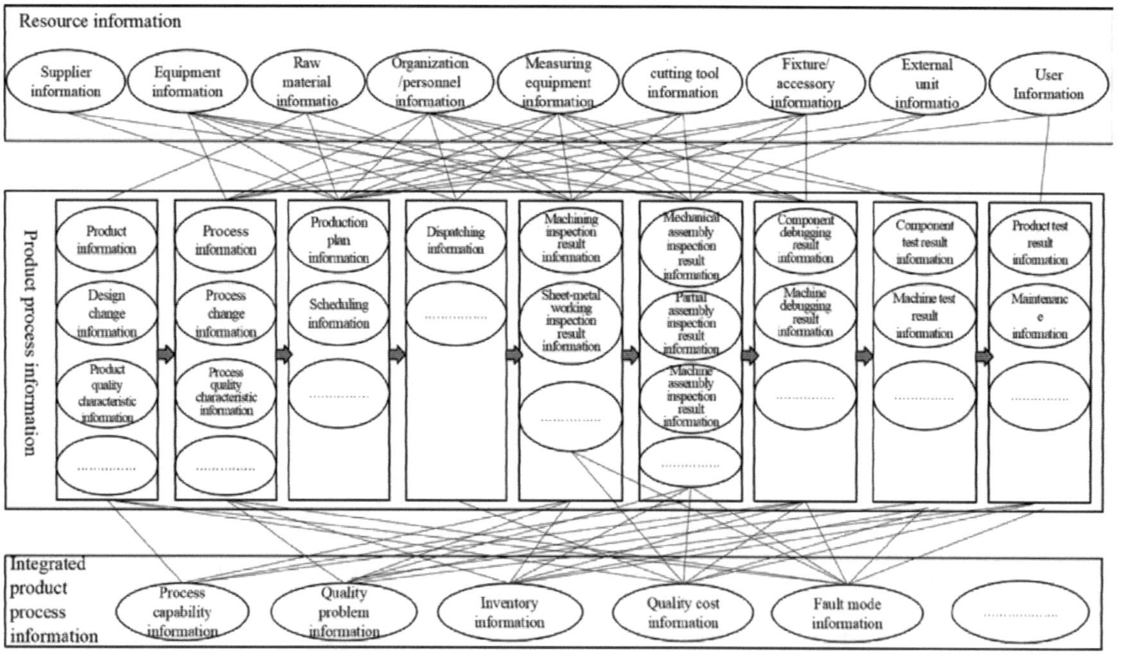

Figure 5. Composition and correlation of quality data

IOP Conf. Series: Materials Science and Engineering **565** (2019) 012002 doi:10.1088/1757-899X/565/1/012002

This paper analyzes the environment for the production of the quality data from the perspective of manufacturing process operation management above and analyzes how to evaluate the production plan, logistics, inventory and manufacturing process through quality data, as well as the correlation between quality data. At the same time, the quality data is formed at various stages of the product life cycle. Therefore, we need to analyze the management category of quality big data from the perspective of product life cycle.

Viewing quality big data from the perspective of product life cycle, the quality data can be classified into three categories: basic resource information, product process information and product process comprehensive information. The main composition and correlation of various types of information are shown in Figure 5.

The management category of quality big data includes the stage information of product formation process and the operation management information of each stage. It is a true multi-dimensional, multi-data type and strongly correlated data set.

3. Acquisition of quality big data

3.1. Purpose of acquisition of quality big data

The quality big data is acquired to effectively monitor and quantitatively assess the production operation, equipment maintenance operation, quality operation situation and inventory operation situation in the manufacturing process, then find out the bottleneck problems and optimize the manufacturing process, to achieve the purpose of gradually improving the manufacturing management system. From the perspective of product quality, process quality and work quality, the specific purpose of acquisition of the quality big data is shown in Figure 6:

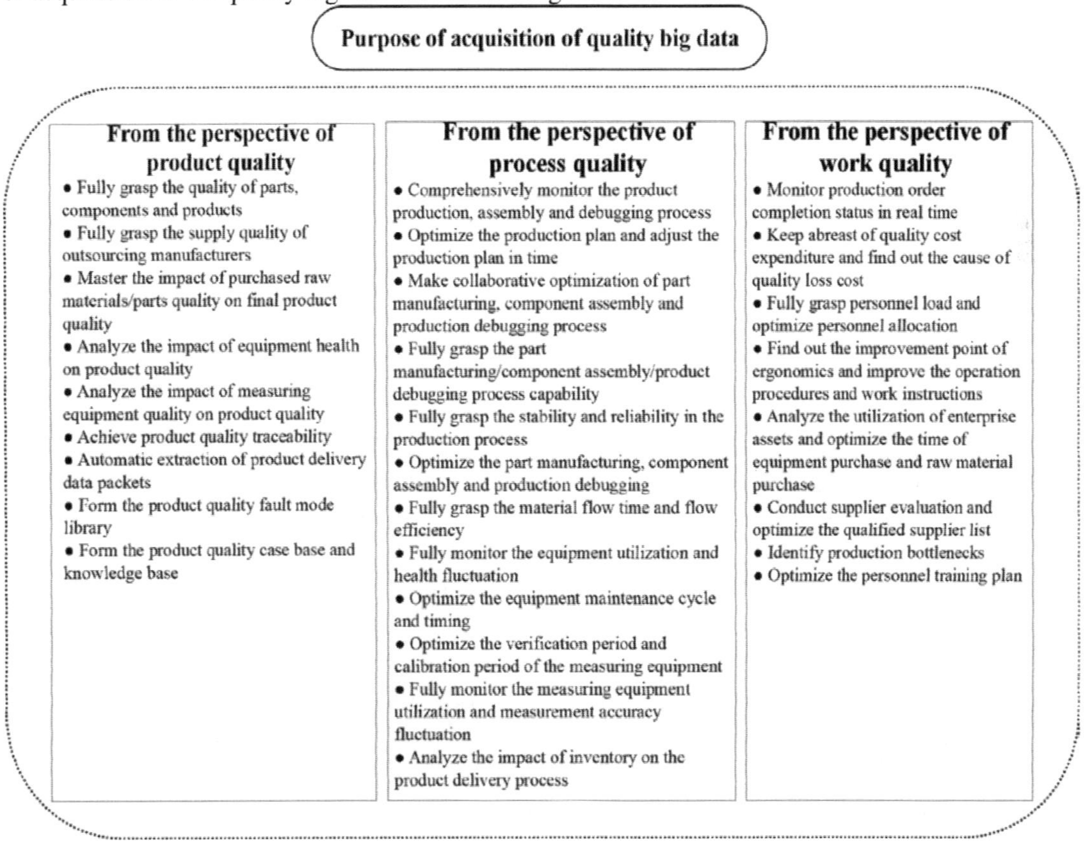

Figure 6. Purpose of acquisition of quality big data

3.2. Acquisition strategy of quality big data

Quality data includes data from process dimension, phase dimension and resource dimension and includes input information such as process and characteristic definitions and production plans, as well as output information of process execution results. Take the acquisition of quality data of a workshop production process as an example. Figure 7 is an example of the input data and output data of the manufacturing process.

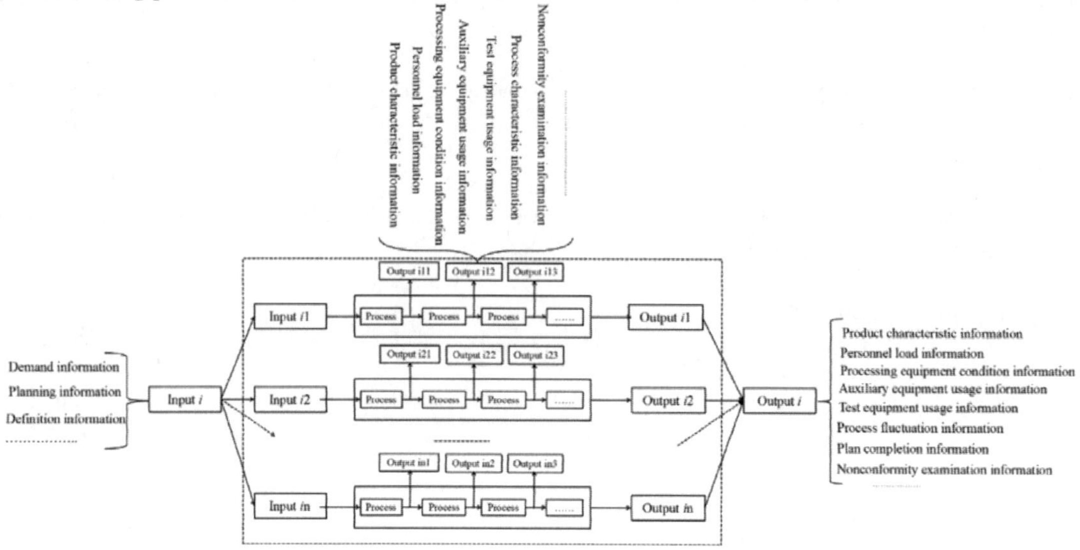

Figure 7. Sample of quality data acquisition during production

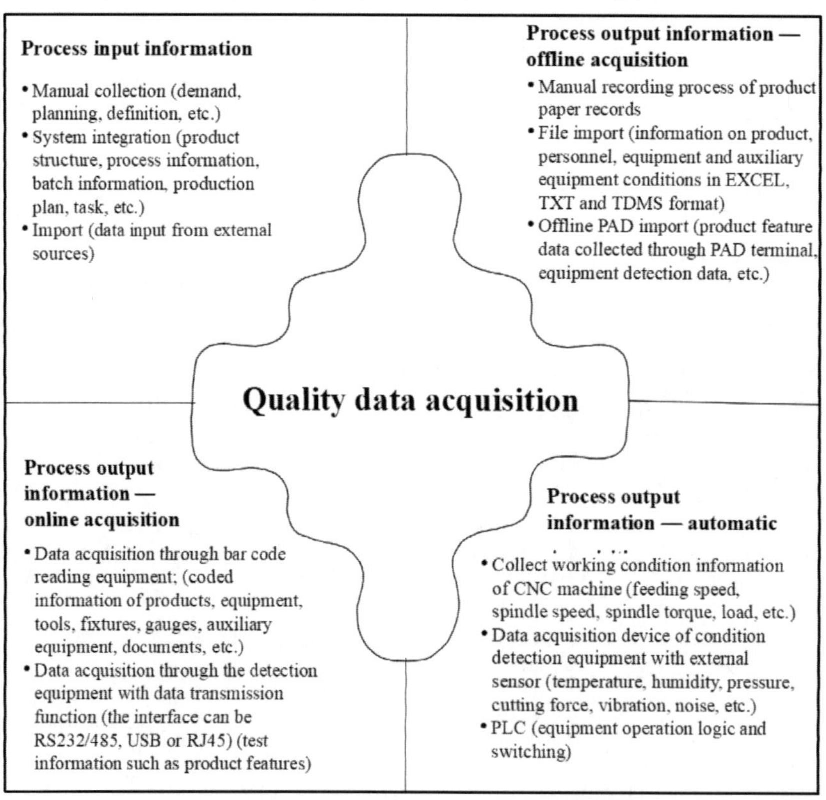

Figure 8. Quality data acquisition method

These data can be acquired both manually and automatically. The manual mode also includes discrete acquisition and online acquisition. Discrete manual quality data acquisition can be recorded by means of informatization of paper records, file import, offline PDA, etc. The manual online acquisition is to use the test equipment with data transmission function to realize the acquisition of test results data triggered manually. Automatic data acquisition is to automatically collect the working condition data of the production process through the data acquisition card in the specified sampling period by adding a sensor. The data acquisition method is shown in Figure 8.

3.3. Acquisition process of quality big data

The quality data is collected and integrated according to the logical level of unit - workshop - enterprise. The acquisition process is shown in Figure 9.

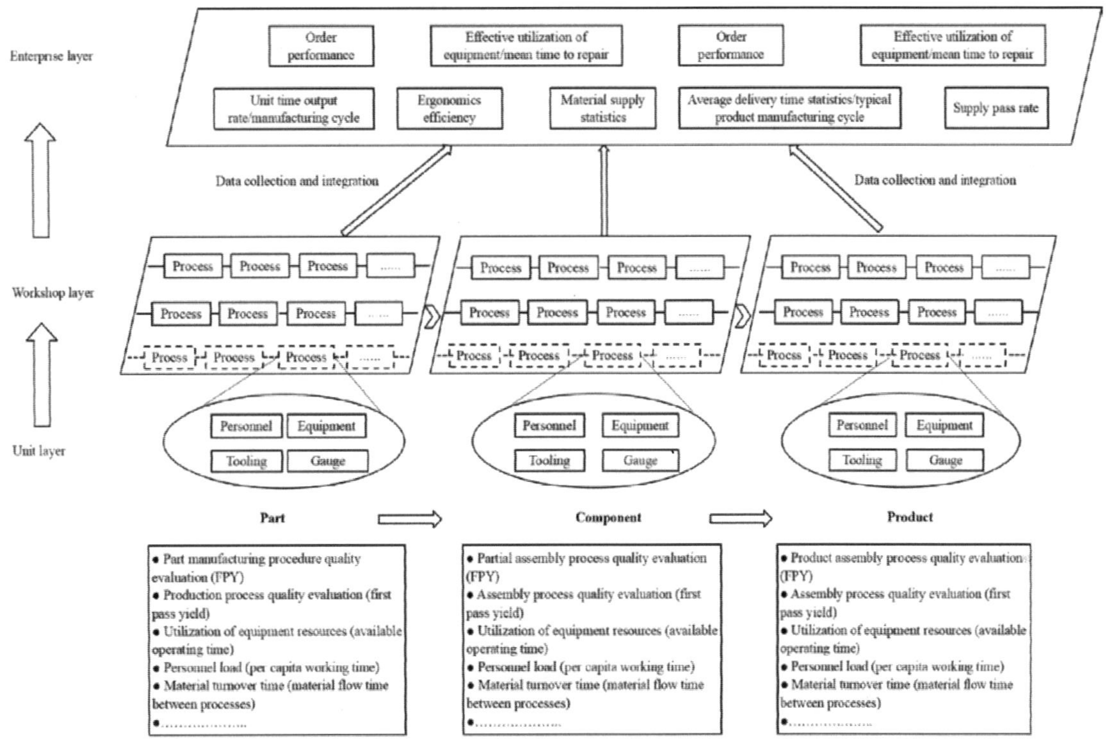

Figure 9. Flowchart of quality big data acquisition

4. Storage specification of quality big data

4.1. Based on unified coding specification

For storage and analysis of quality data, the quality data shall be stored based on unified basic resource coding and material coding.

For the basic resource coding, full consideration shall be taken into the management requirements of classification, stratification and uniqueness, and the compilation of unified basic coding for enterprise resource sharing. The management basis is provided for realizing unified standard management of basic resources, monitoring of resource status, uniform optimized allocation of resources, and forecasting of resource purchase quantity and time. Basic resources include: equipment, tools, measuring tools, fixtures, auxiliary equipment and personnel. The unified coding of basic resources is the basis and key to realize the quality management of production site based on Internet of Things.

The material coding shall take full account of the requirements for classification and stratification in management and for the correlation with the processing of products and parts in the production process, and is the basis and key to achieve the traceability of materials for products and production processes.

4.2. Based on business flow and data flow framework

The storage of quality big data shall match the actual business flow, and a detailed business flow and data flow framework shall be developed, as shown in Figure 10. The framework implements a top-down demand planning, task decomposition, manufacturing execution and field control and bottom-up information feedback and achieves the quality analysis and evaluation of the products, equipment, materials, personnel, production plan, manufacturing process and supplier through input and output analysis and comparison of each unit and process.

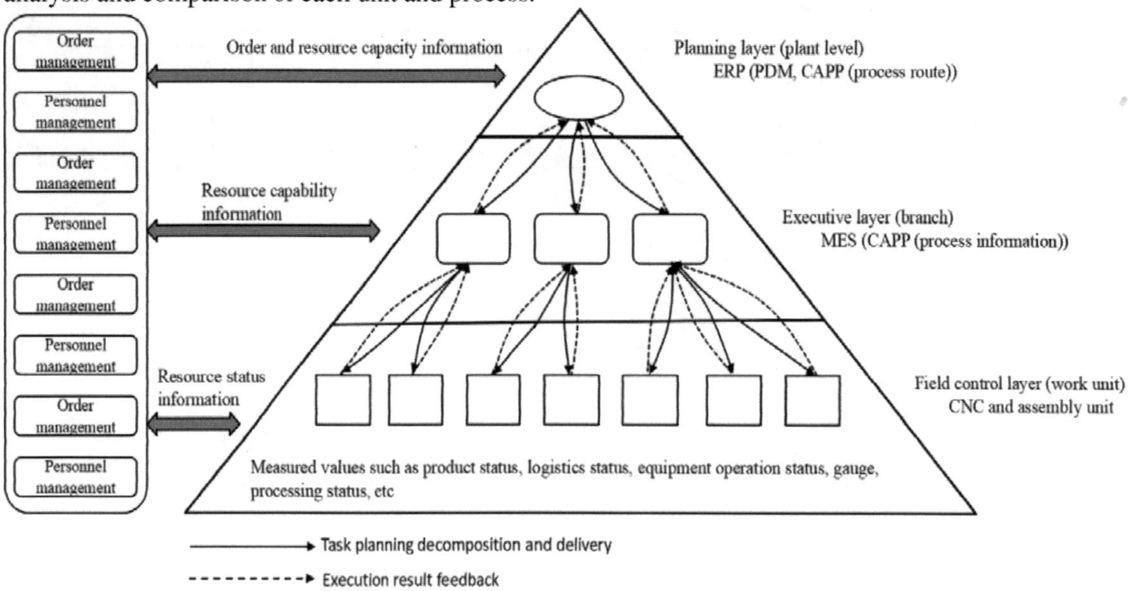

Figure 10. Business flow and data flow framework of quality big data management system

4.3. Quality data modeling and expression specification

All quality data are output of inspection or measurement activities, which are generated in a certain situation. Only the quality data with detailed situation information can be used for the quality problem tracing and the statistics and analysis of process, production process, batch and product. At the same time, the optimization of production process, of personnel workload and of production task can be realized by calculating operation time, process waiting time and inspection time through the start time and completion time of each process. It can be seen that the situation information of quality data is the basis for quality data classification, automatic collection, quality data based statistical analysis and process optimization. Therefore, it is necessary to establish the expression specification of quality data. The organization model and components of the quality data are shown in Figure 11.

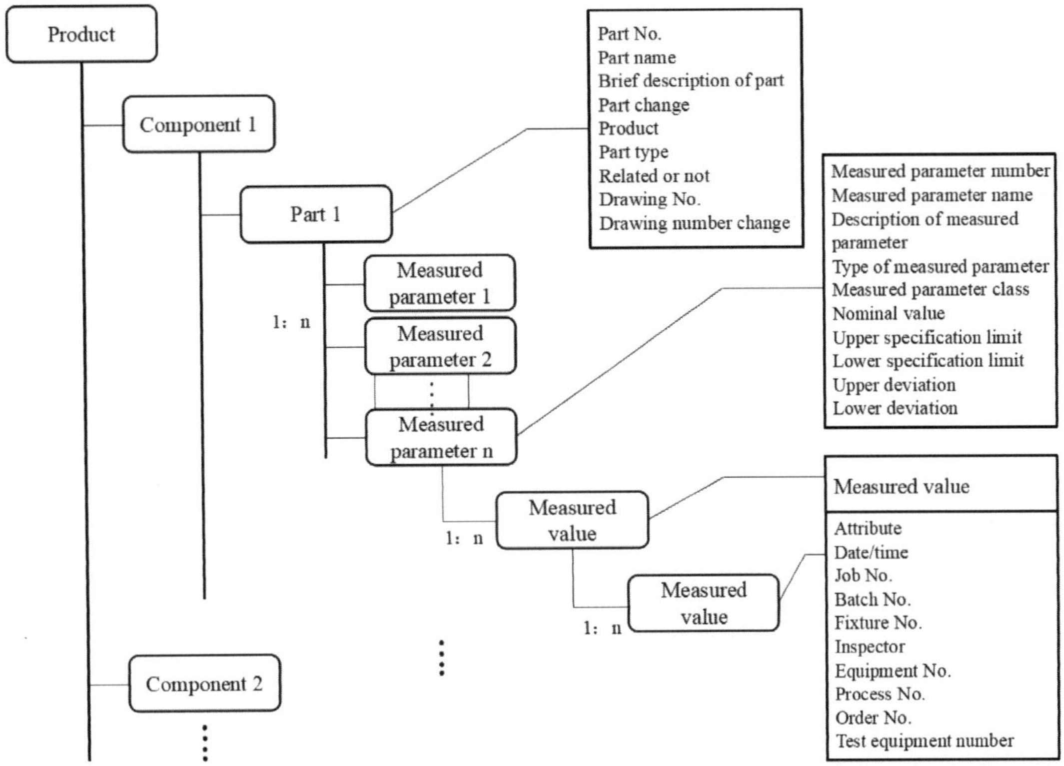

Figure 11. Organization model and components of quality data

5. Conclusion

Data integrity and availability is an important aspect of data acquisition and data management. Only corresponding to the product, parameter name, measurement time, surveyor and the product production equipment correctly, a quality data can truly play the value, provide material available for data analysis in the future and provide data support for enterprises to improve product quality, process quality and work quality.

References

[1] Li J F, Shu Q L and Hao B 2004 *J. Shenyang Inst. Technol.* **23** 11
[2] Wu Y F Study on factory digitization practice based on MES system 2016 Yunnan University of Finance and Economics
[3] Liu Y Analysis and evaluation research of lean management based on execution 2012 Xi'an University of Science and Technology
[4] Li Z C Research and development of meticulous production management system for mould enterprises 2012 Hefei University of Technology
[5] Tian N The implementation of Lean quality management research of small and medium-sized enterprises 2012 Wuhan University of Technology

IOP Conf. Series: Materials Science and Engineering **565** (2019) 012003 doi:10.1088/1757-899X/565/1/012003

Application of Particle Swarm Optimization BP Algorithm in Air Humidity of Greenhouse Crops

Guo Li-Li[1], Liu Yong[1*], Liu Zi-Wei[1] and Sun Hua-Wei[2]

[1]College of Electronic Engineering, Heilongjiang University, Harbin 150080, China
[2]Heilongjiang Eastern Water Saving Equipment Co. Ltd., Suihua 150000, China

[*]E-mail: liuyong@hlju.edu.cn

Abstract: The advancement of artificial intelligence, the synonym of precision agriculture has approached the public's vision, and the different requirements for air humidity in different growth periods of crops are proposed. The BP neural network optimized by particle swarm optimization algorithm is proposed to predict the air humidity of crops. Algorithm, this paper chooses BP algorithm network topology structure is 2-5-1, improves the inertia weight of PSO algorithm, proposes nonlinear inertia weight reduction strategy $w = w_s - (w_s - w_e)\left|f_1\sqrt{\frac{t}{T}}\right|$, trains BP algorithm with improved PSO algorithm, has no gradient information, jumps out local pole Value, reduce the number of iterations of the algorithm, and speed up the training speed of the neural network. According to the experimental results of MATLAB simulation, the air humidity prediction model of particle swarm optimization neural network is constructed, which proves the effectiveness of the improved particle swarm neural network prediction system. It can be shown that the proposed algorithm has a relative error of at least 0.0134 compared with other algorithms.

Key words: intelligent agriculture; air humidity; BP neural network; particle swarm optimization

1. Introduction

This paper proposes to combine the improved particle swarm optimization algorithm with BP neural network and apply it to the prediction of air humidity required by crops. In 1986, Rumelhart et al. [1] proposed the BP neural network algorithm, which solved the problem of hidden unit connection weight in multi-layer neural networks and has been widely used. However, BP algorithm based on gradient descent is easy to fall into local poles. Inadequate defects such as value and slow convergence rate. In 1995, the particle swarm optimization algorithm PSO proposed by American scholars Eberhart and Kennedy [2] introduced them into the training of neural networks and received wide attention.Song Xiaomeng et al [3] introduced the Internet of Things technology and fuzzy PID control to In the greenhouse greenhouse environmental monitoring system, from the aspects of low cost, precise control and easy operation, the intelligent monitoring of strawberry greenhouses is realized. Wang Yaodong [4] proposed to construct a seismic prediction model based on improved particle swarm neural network, using inertia. Weights and learning factors dynamically adjust the strategy balance algorithm, and get a lower average error rate; Zhang Ermei [5] based on the characteristics of ammunition storage reliability, established a prediction model combining particle swarm and neural network, which added the momentum term. And variable learning rate parameters, to achieve the function of ammunition storage reliability assessment and prediction.

In this paper, the inertia weights in the PSO algorithm are improved, and combined with the BP algorithm in the prediction of crop air humidity, the purpose is to control the air humidity demand of different growth periods of crops through the effective prediction of intelligent algorithms. Bakker [6] Studies have shown that changes in humidity in the greenhouse can affect the yield of growing crops. In the case of high humidity, crops are easily corroded by fungal diseases, directly affecting the quality and yield of crops. On the contrary, when the humidity is too low, It will cause serious water loss of crops and lead to poor growth or even death of crops. Therefore, effective control of air humidity is necessary, and the air temperature and humidity in the greenhouse have a certain coupling relationship. We can stabilize the air temperature. Apply BP intelligent algorithms to air humidity.

2. Improved particle swarm optimization algorithm

The PSO particle swarm optimization algorithm has a fast calculation speed and the algorithm itself has few parameters and strong global search ability. In order to improve the convergence speed and accuracy of the algorithm, a nonlinear weight decrement strategy is proposed in this paper. The experimental results show that the effect is the best [7] .

In order to balance the algorithm's global search and local search ability, and to solve the shortcomings of the algorithm's low convergence and falling into local extremum, in 1998, YHShi et al. introduced the inertia weight into the PSO algorithm, and improved the particle swarm algorithm. To ensure a good convergence effect, the evolution process is shown in equations (1) and (2):

$$V_{ij}(t+1) = wV_{ij}(t) + c_1 r_1 [V_{ij}(t) - X_{ij}(t)] + c_2 r_2 [P_{gj}(t) - X_{ij}(t)] \quad (1)$$

$$X_{ij}(t+1) = X_{ij}(t) + V_{ij}(t+1) \quad (2)$$

Where i=1, 2...n, j=1, 2...d, n represents the number of particles in the particle group, and d represents the dimension of the target space. c_1 and c_2 represent learning factors, r_1 and r_2 are random numbers, the value range is $[0,1]$, and w represents inertia weight.

The standard particle swarm optimization algorithm is commonly used in Y.H.Shi[8] to propose a linear decrement weight strategy expression as shown in (3):

$$w = w_s - (w_s - w_e)\frac{t}{T_{\max}} \quad (3)$$

The search process is a nonlinear and complex process, so this paper improves the linear decreasing weight strategy into a nonlinear decreasing weight strategy to enhance the local and global search ability. The formula is shown in (4):

$$w = w_s - (w_s - w_e)\left| f_1 \sqrt{\frac{t}{T_{\max}}} \right| \quad (4)$$

Where $T_{\max}$ represents the maximum evolution algebra; w_s represents the initial inertia weight value, w_e represents the inertia weight that evolved to the maximum allowed number of iterations, t represents the current number of iterations, f_1 is the control factor, and $w_s = 0.95$, $w_e = 0.4$. When $t = 0$, the function value is equal to the initial weight w_s . When $t = T_{\max}$, different values of f_1 will get different decreasing effects. When f_1 is larger, it will decrease to close in the early stage of the algorithm. Forcing the algorithm to fall into the local search early; when the value is small, it will make w more than we when $t = T_{\max}$. When the experimental test f_1 takes the value of 0.9, the experimental effect is the best, so the value of f1 in this paper is 0.9. The result shows that the improved strategy can better maintain the balance between global search and local search.

3. Particle Swarm Optimization of BP Neural Network

The particle swarm optimization algorithm optimizes the initial weight and threshold of the BP neural network, and sets the position in the particle swarm as the set of weights and threshold values in the current iteration in the BP network, ie $p_g = \{W_1 B_1 W_2 B_2\}$, where W_1 is the hidden layer weight matrix, $W_1 = \begin{bmatrix} w_{31} & w_{32} \\ w_{41} & w_{42} \\ w_{51} & w_{52} \\ w_{61} & w_{62} \\ w_{71} & w_{72} \end{bmatrix}$, W_2 For the output layer weight matrix, $W_2 = \begin{bmatrix} w_{83} & w_{84} & w_{85} & w_{86} & w_{87} \end{bmatrix}^T$,

B_1 is the hidden layer threshold matrix, $B_1 = \begin{bmatrix} \theta_3 & \theta_4 & \theta_5 & \theta_6 & \theta_7 \end{bmatrix}^T$, B_2 is the output layer threshold matrix, and the dimension of the $B_2 = \begin{bmatrix} \theta_8 \end{bmatrix}$ particle is mapped to the number of weights and the number of thresholds that are connected, through the optimal position The optimization process obtains the initial weight and threshold of the optimal BP neural network, making full use of the global search ability of the PSO algorithm and the local search ability of the BP neural network.

3.1 Particle Swarm Optimization Algorithm for BP Neural Network Optimization

(1) Initialization parameters: such as BP neural network topology, particle numbe $n = 30$ r, particle dimension $d = 21$, maximum iteration number $T_{\max} = 1000$, learning factor $c_1 = c_2 = 2$, inertia weight $w_s = 0.95$, $w_e = 0.4$, learning rate $lr = 0.05$, target error $goal = 0.0001$,maximum particle velocity $v_{\max} = 1$ and minimum speed $v_{\min} = -1$ and other parameters;

(2) Calculating the fitness value of the particle, that is, the neural network output error, determining the individual extremum and the global extremum of the particle: for each particle i, the fitness value p_i is compared with the individual optimal value $p_i^{'}$, if $p_i < p_i^{'}$, Then $p_i^{'} = p_i$, and record the position of the current best particle; for each particle i, compare the fitness value p_i with the global optimal value p_g, if $p_i < p_g$, then $p_g = p_i$, and record the position of the current best particle;

(3) Updating the velocity and position of each particle according to formulas (1) and (2), and judging whether the speed and position of the updated particle are within a limited range;

(4) Checking the iterative stop condition of the algorithm. If the result is ideal, stop the iterative output of the BP neural network weight or threshold, so that the BP network can better reduce the error, otherwise go to step (2).

3.2 Particle Swarm Optimization Algorithm and Neural Network Simulation Analysis

In this experiment, the cultivation of strawberries in greenhouses in Suihua City, Heilongjiang Province was taken as an example to study the air humidity requirements of strawberries in different growth periods in greenhouses. Strawberry humidity control requirements in greenhouse: The relative air relative humidity required for strawberry flowering is 40%; the initial temperature of strawberry is to prevent strawberry from entering dormancy period, and the required temperature is relatively high. At this time, the relative humidity of air is controlled at 85%-90%; The flowering period of strawberry is stricter to the air humidity. If the humidity is too small or too small, the pollination will be poor. At this time, the relative humidity of the air is controlled at about 40%; the fruit of the strawberry is swollen and mature, and the strawberry is affected by the temperature at this time. The humidity can be controlled between 60% and 70%.

According to these control requirements, this paper uses an optimized intelligent algorithm to control and construct a 3-layer BP neural network applied to the prediction of greenhouse air humidity, in which there are 2 input layer nodes and 5 hidden layer nodes, hidden layer. The activation function is 'tansig', the output layer node is one, and the output layer activation function is 'logsig'. In the application, the two parameters of sun light and air temperature in the shed are taken as input variables, and the optimum air humidity required by strawberry is used as the output variable for simulation test. This simulation experiment selects 30 data of winter strawberry, the first 20 data. For the training data, the last 10 data are the detection data. In order to better reflect the excellent value of the result, we compare the results of BP network algorithm, linear PSO-BP network algorithm and nonlinear PSO-BP network algorithm. The simulation results are as follows: The Figure shows:

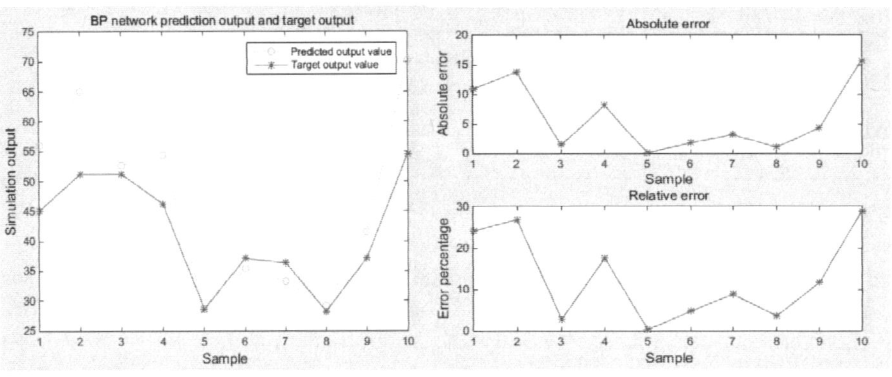

Figure 1. BP network prediction output and target output and associated error map

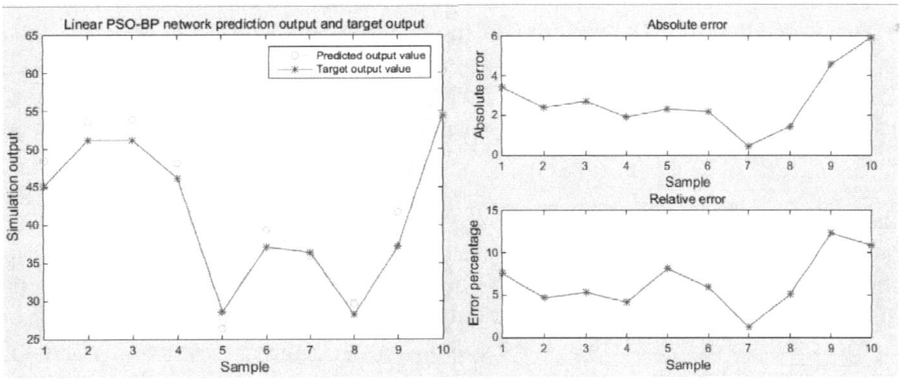

Figure 2. Linear PSO-BP network prediction output and target output and associated error map

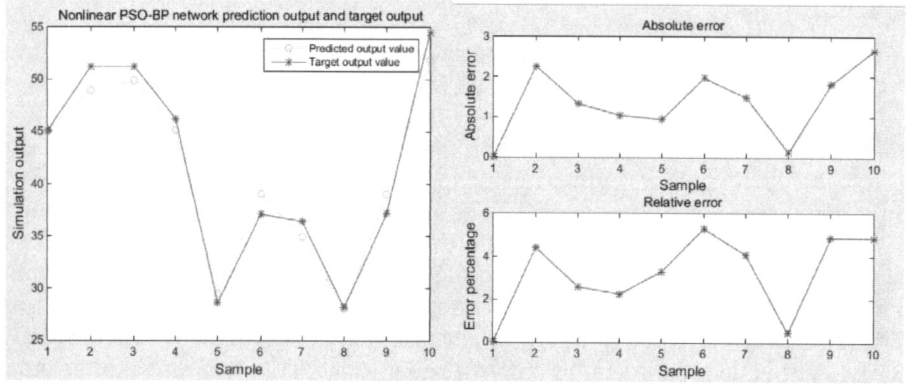

Figure 3. Nonlinear PSO-BP network prediction output and target output and associated error map

Table 1. Comparison of average error rates of the three algorithms

Algorithm selection	Relative error 21	22	23	24	25	26	27	28	29	30	Average error rate
BP	24.1901	26.8582	2.7993	17.6296	0.3141	4.7043	8.8011	3.5698	11.7018	28.7990	12.9367
LinearPSO-BP	7.5760	4.6593	5.2843	4.1138	8.0835	1.5592	1.2038	5.0751	12.3016	10.8968	6.0753
NolinearPSO-BP	0.0134	4.4126	2.5937	2.2553	3.3092	5.3133	4.0995	0.4464	4.8785	4.8665	3.2188

Result analysis:

It can be seen from the above simulation results that the BP neural network optimized by nonlinear PSO is much smaller than the other two algorithms in training error. Especially for the traditional BP network algorithm, the training effect is much better than the traditional BP network. The linear PSO-BP algorithm has strong learning ability and generalization ability. It can be seen from Figure. 1 that the traditional BP is easy to fall into the extreme value, especially the target output value of the sample 22 reaches 64.9, far exceeding the predicted output value, and the error. The rate is also the highest of the three algorithms; from Figure. 2 and Figure 3, the linear PSO-BP network and the nonlinear PSO-BP network have similar training effects, but the nonlinear PSO-BP network has a linear error rate. Smaller, the absolute error of the nonlinear PSO-BP network is less than 3, and the error is minimum at 0.0060 when the sample is 21, which shows the best training effect. Finally, the nonlinear PSO-BP network can be seen from Table 1. The relative error is the smallest, the accuracy is the highest, and the global search ability is also good. In summary, the optimization scheme proposed in this paper is feasible.

4. Conclusion

Aiming at the prediction of air humidity in agricultural planting, this paper proposes an intelligent optimization algorithm based on particle swarm optimization PSO to optimize BP neural network. The algorithm combines the idea of population activity mechanism and back propagation neural network, which improves the global algorithm to some extent. Search ability, convergence accuracy and speed, so as to quickly control the demand for air humidity in different growth periods of crops, so that crops have a better growing environment.

References

[1] Rumelhart D E, Hinton G E, Wliiians R J. Learning representations by back-propagation error. Nature, 1986, 323(9): 533-536.
[2] Kennedy J, Eberhart R C. Particle swarm optimization. IEEE International Conference on Neural Networks, 1995: 1942-1948.
[3] Song Xiaomeng. Research and development of intelligent monitoring system for strawberry greenhouses [D]. Shandong Agricultural University, 2018.

[4] Wang Yaodong. Optimization and application of BP neural network based on improved particle swarm optimization algorithm [D]. Xi'an University of Science and Technology, 2012.
[5] Zhang Ermei. Research on the reliability prediction of ammunition storage based on BP neural network based on particle swarm optimization[D]. Shenyang Ligong University, 2018.
[6] Korner O, Challa H. Design for an improve temperature integration concept in greenhousecultivation [J]. Computers and electronic in agriculture, 2003, 39(1): 39-59
[7] Tian Yubo. Hybrid Neural Network Technology. Beijing: Science Press, 2014.
[8] Shi Y,Eberhart R C.Empirical study of particle swarm optimkization [C].Proc of Congress on Computational Intelligence,Washington DC,USA,1999:1945-1950.

IOP Conf. Series: Materials Science and Engineering **565** (2019) 012004 doi:10.1088/1757-899X/565/1/012004

Research and Application of Bottleneck-based Planning and Scheduling Method for PCB Manufacturing System

Dan Luo, Zailin Guan, Cong He, Weikang Fang, Chuangjian Wang* and Zhangyi Zeng

State Key Lab of Digital Manufacturing Equipment and Technology, School of Mechanical Science and Engineering, Huazhong University of Science and Technology, Wuhan 430074, China

*Email: wangchj@wust.edu.cn

Abstract. Researches on planning and scheduling of PCB production system are gaining more and more attention due to the complexity and uncertainty of it. Current research focused on the solving method how to make better schedule solution for PCB system based on a real case of PCB production enterprise. Firstly, for the bottleneck process of electroplating, a mixed integer programming model is proposed, moreover, a branch and bound method is employed to solve the problem. Secondly, for non-critical processes, the man-hour quota is predicted using the neural network model and the association rules, and then the data can be found using the association rules and be modified. Thirdly, the planning and scheduling method based on bottleneck is used for the order sequencing, and "pull-push" mechanism of Drum-Buffer-Rope are applied to generate the scheduling scheme for non-bottlenecks. In the end, the effectiveness of the proposed scheduling method is proved by computational experiments.

1. Introduction

With the rapid development of electronic communication technology, the market demand of printed circuit board (PCB) is greatly increasing. Due to the complexity of PCB production process, the variety of products and the uncertainty factors, it is very difficult to make production plan for PCB production. With the intensification of competition, more and more attention has been paid to the formulation of production plan for PCB manufacturing system.

There are many literatures about scheduling problem in PCB manufacturing system. However, most studies only focus on scheduling of single processor or workstation, such as sorting scheduling of surface mounting equipment [1] and grabbing scheduling of electroplating workstation [2]. Very few researches are about the overall production schedule of PCB manufacturing system. Kim et al. [3] proposed a heuristic algorithm for PCB assembly systems, but their research is limited to typical

IOP Conf. Series: Materials Science and Engineering **565** (2019) 012004 doi:10.1088/1757-899X/565/1/012004

(non-hybrid) flow shop scheduling problems. Referring to the semiconductor wafer planning method, PCBs planning can use or modify various scheduling methods or rules, such as Dessouky and Leachman [4], Kim [5] and Dabbas [6].

In summary, it is of great significance both in academia and industry to study the production plan of PCB workshop. This paper is devoted to optimizing the bottleneck production planning schedule and forming the overall production plan of PCB workshop, so as to achieve the goal of improving the productivity and efficiency of PCB workshop.

2. Description

There are various logic of series, parallel and reflux in the process flow of PCB workshop. If all processes are considered in a same production scheduling level, as a result, the scheduling model will be super big with high computational complexity. Therefore, several key processes are selected out, and only the sequence of critical processes is carried out in production scheduling. According to historical data, automatic optic inspection (AOI), oxide replacement, lamination, drilling, electroplating, welding resistance and final inspection are selected as critical processes. In the critical process, electroplating is the bottleneck. The processes which are not belong to critical process are non-critical processes.

In current research, bottleneck process problem, i.e. electroplating scheduling problem, is summarized as a parallel machine scheduling problem with shared hoists, which can be summarized as follows. Consider the problem that N tasks are processed on M parallel electroplating tanks. M parallel electroplating tanks share the same hoists, as shown in figure 1. This paper has the following assumptions: (1) N independent workstations are non-preemptive and are processed in parallel processor units; (2) The operation cycle of the hoist is T, the unloading time is T, and the loading time is 0, but the loading time is T when continuous loading; (3) All the hoist and electroplating tanks are idle at 0 time; (4) The shared hoist and the electroplating tanks can only handle one task at a time; (5) The time for the hoist to each electroplating tank is the same; (6) The hoist and electroplating tanks will not break down.

In order to avoid the tedious and lengthy formulation of production plans, the man-hour quota is used instead of the standard processing time to calculate the processing time of production in non-critical processes. It is called the man-hour quota of the process from the entry of the task to the completion of the processing of the order. It includes the basic processing time, auxiliary time, layout time, rest time and so on, as shown in Figure 1.

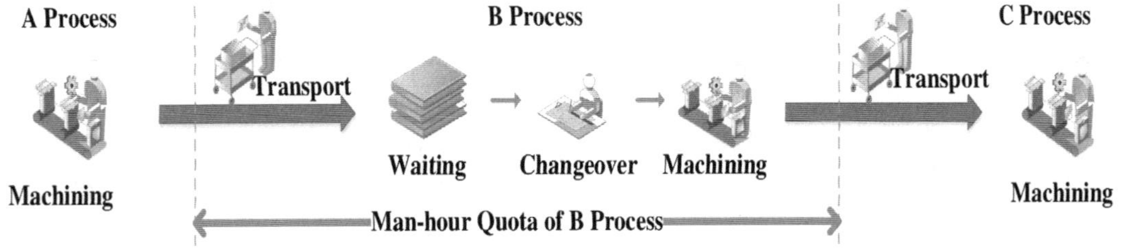

Figure 1. Diagram of man-hour quota

IOP Conf. Series: Materials Science and Engineering **565** (2019) 012004 doi:10.1088/1757-899X/565/1/012004

3. Production planning method

3.1 Bottleneck process scheduling optimization based on branch and bound method

3.1.1 Model parameters and constraints. The basic parameters are shown as following.

N	Number of work tasks
M	Number of parallel electroplating tank
T	Cycle time of shared hoist
PT_i	Processing time of working task i
SH_k	Starting time of the kth activity of the hoist, $k = 1,2,3,\ldots,2n$
$Cmax$	Completion Time
Ω	Maximal value

Decision variables:

$$x_{ijk} = \begin{cases} 1, & \text{If the } k\text{th activity of hoist is to load task } i \text{ into tank } j \\ 0, & \text{otherwise} \end{cases}$$

$$y_{ijk} = \begin{cases} 1, & \text{If the } k\text{th activity of hoist is to unload task } i \text{ from tank } j \\ 0, & \text{otherwise} \end{cases}$$

Objective function and constraints:

$$Min\ Z = C_{max} \tag{1}$$

$$s\ t:$$

$$C_{max} \geq SH_{2N} + T \tag{2}$$

$$\sum_{j=1}^{M}\sum_{k=1}^{2N} x_{ijk} = 1;\ \forall i = 1, 2, \ldots, N \tag{3}$$

$$\sum_{j=1}^{M}\sum_{k=1}^{2N} y_{ijk} = 1;\ \forall i = 1, 2, \ldots, N \tag{4}$$

$$\sum_{j=1}^{M}\sum_{i=1}^{N} (x_{ijk} + y_{ijk}) = 1;\ \forall k = 1, 2, \ldots, 2N \tag{5}$$

$$\sum_{k=1}^{2N} x_{ijk} = \sum_{k=1}^{2N} y_{ijk};\ \forall i = 1, 2, \ldots, N,\ \forall j = 1, 2, \ldots, M \tag{6}$$

$$\sum_{k=1}^{k'}\sum_{j=1}^{M} x_{ijk} \geq \sum_{k=1}^{k'}\sum_{j=1}^{M} y_{ijk};\ \forall i = 1, 2, \ldots, N,\ \forall k' = 1, 2, \ldots, 2N \tag{7}$$

$$SH_{k+1} \geq SH_k + \sum_{j=1}^{M}\sum_{i=1}^{N}\sum_{j'=1}^{M}\sum_{i'=1}^{N}\left[T \cdot y_{i'j',(k+1)} + T \cdot x_{ijk} . x_{i'j',(k+1)}\right];\ \forall k = 1, 2, \ldots \tag{8}$$

$$SH_{k+1} \geq SH_l + \sum_{i=1}^{N}\left[\left(\sum_{j'=1}^{M}\sum_{i'=1}^{N} T \cdot x_{i'j',(l-1)} + PT_i\right) \cdot x_{ijl} + T \cdot y_{ijl}\right] + \Omega\left[\sum_{i=1}^{N}\left(x_{ijl} + y_{ijl} + x_{ij,(k+1)}\right) - 2\right]; \ \forall k = 1, 2, \ldots, 2N-1, \ \forall l = 1, 2, \ldots, k, \ \forall \tag{9}$$

$$x_{ijk} \in \{0, 1\}; \ y_{ijk} \in \{0, 1\}; \ \forall i = 1, 2, \ldots, N, \ \forall j = 1, 2, \ldots, M, \ \forall k = 1, 2, \ldots, 2N \tag{10}$$

The model takes the activities of the hoist as decision variables, and N tasks correspond to $2N$ activities of the hoist. Equation (1) indicates that the model aims at minimizing the completion time. Constraint (2) is about the completion time and the hoist. Constraint (3) and (4) indicate each task is loaded and unloaded only once. Constraint (5) ensures the hoist can only handle one task at a time. Constraint (6) means that the corresponding electroplating tanks are the same for each task loading and unloading. Constraint (7) describes that the task can only be loaded before unloaded. Constraint (8) shows the constraint relationship between two adjacent start times. Constraint (9) means the constraint of the release time of the electroplating tank. Constraint (10) is constraints about decision variables and parameters.

Algorithmic logic of branch and bound:

Step 1: A preliminary scheduling result can be obtained by using MSPT algorithm, and the total completion time of the scheduling is taken as the upper bound (UB). Let $r = 0$, the corresponding $\mathrm{R(r)}$ be empty, and the branching process starts at node 0.

Step 2: Deciding on possible branches. When a node is selected, the unscheduled work J_r' and empty processor can be figured out. If there are empty processor and unscheduled jobs, the number of jobs to be scheduled is the range of possible loading choice (PLC), otherwise PLC=0. If the processor is not empty, the number of jobs on the processor is the range of possible unloading choice (PUC), otherwise PUC = 0. Therefore, the possible branch (PB) of a node is equal to the sum of PLC and PUC. Let $\mathrm{R(r+1)} = \mathrm{R(r)}$ and add 1 in turn.

Step 3: Calculate the lower limit. The earliest starting time of processor is calculated by REST algorithm. STPT rule and the m-jobs-at-a-time assignment procedure can be used to determine the scheduling order. The lower limit of completion time is calculated by formula (1) and (2). Repeat until all branches are computed. Compare the lower limit of each node with UB and eliminate the remaining nodes. If $\mathrm{r} = 2\mathrm{n} - 1$, go to step 5, otherwise go to step 4.

Step 4: Select branches. Select a node with a minimum lower bound, update its corresponding r and $\mathrm{R(r)}$ values, and go back to step 2.

Step 5: Measurement process. UB is replaced by the value of the new solution. Other nodes with lower bounds greater than or equal to UB will be deleted. If all the nodes have been measured, an optimal solution will be obtained; otherwise, return to step 4.

3.2 Prediction and analysis of man-hour quota of non-critical processes based on BP neural network and association rule analysis

This paper analyses the man-hour quota of non-critical processes in PCB workshop by using big data, data mining and other technologies. First, the BP neural network prediction model is adopted to preliminarily predict the man-hour quota of orders in different processes. Then the association rules are used to find out the process needed to be revised, and revise it.

3.2.1 Prediction of man-hour quota based on BP neural network model. Back Propagation (BP) network, well-known neural network models nowadays, is a multi-layer feedforward network trained by error back propagation algorithm. Because the influencing factors and weights of different processes are different, it is necessary to establish different prediction models for different production processes. Current study takes electronic testing as an example to describe the process of building neural network prediction model. The BP neural network model of man-hour quota prediction for electronic testing is shown in the Figure 2, which consists of three layers: input layer, hidden layer and output layer. The input layer includes six attributes, such as board type, PNL length. The predicted value of the man-hour quota of the output layer is obtained by processing nine nodes of the hidden layer.

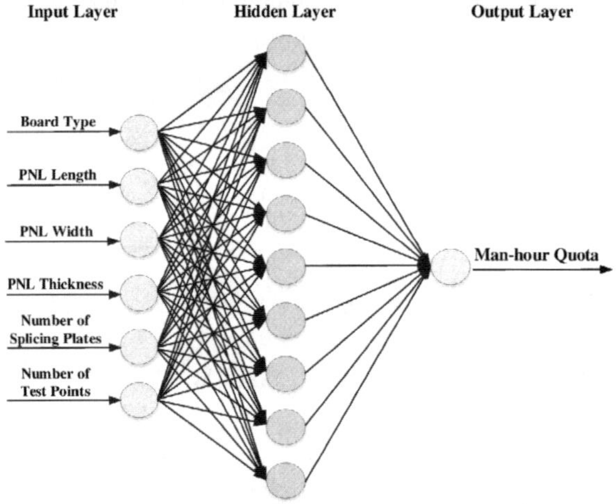

Figure 2. BP Neural Network Model for Forecasting Man-hour Quota

3.2.2 Modify of prediction results based on association rule Apriori algorithm. The steps of Apriori algorithm are as follows:

Step 1: Find frequent item sets I. By counting the occurrence times of all elements in I, the item sets less than the minimum support count threshold min_support are screened out, and the remaining frequent item sets are recorded as set L1.

Step 2: Connection. If two frequent $k-1$ item sets (I1, I2) can be connected, a new k item set is generated from them. All sets that may be composed of frequent item sets k are denoted as C_k.

Step 3: Pruning. To screen C_k, if all k-1 subsets of each candidate item set k are preserved in L_{k-1}, otherwise deleted. The frequency of all item sets in C_k is counted, and the set L_k is obtained by deleting those less than min_support.

Step 4: Repeat step 2 or 3 until find k that $C_k = \emptyset$, obtain all frequent item sets.

This paper uses Python language to implement Apriori algorithm. The parameters in the algorithm are: minimum support is 0.3, minimum confidence is 0.5. After running, all association rules with 1 consequent are obtained. According to the obtained association rules, the records that need to be revised are found. The revise method is as follows:

$$T_r = T + a_i \times (\bar{T} - T)\#(11)$$

T_r is the revised man-hour quota, T is the original man-hour quota, $\bar{T}$ is the average value of the predicted man-hour quota that satisfies the association rule in all records, a_i is the confidence under the association rule.

3.3 Bottleneck-based integrated production scheduling method for PCB workshop

In the current research, we use Hibernate framework with Java language and MySQL database to realize the overall scheduling method of PCB workshop. We use Java algorithm and heuristic scheduling rules to drive data and arrange production. The final scheduling results are output in the form of task data.

Firstly, according to the resource load rate, the processes are divided into two categories: critical process and non-critical process. Secondly, the branch and bound algorithm is used to arrange the production of bottleneck process (electroplating process) so as to maximize the utilization of resources, and the order sequence of electroplating process is guaranteed by setting time buffer. Then using heuristic rules, the production plan after the bottleneck process can be deduced forward, and the production plan before the bottleneck process can be deduced backward. The processing time of critical processes is calculated by standard processing time, but non-critical processes' by man-hour quota. Finite capacity scheduling is used for critical processes and infinite capacity scheduling for non-critical processes. Through the above steps, we can get the planned starting time and ending time of each process of all orders, and the processing equipment of each critical process of orders. To sum up, the whole process of production scheduling is completed, as shown in Figure 3.

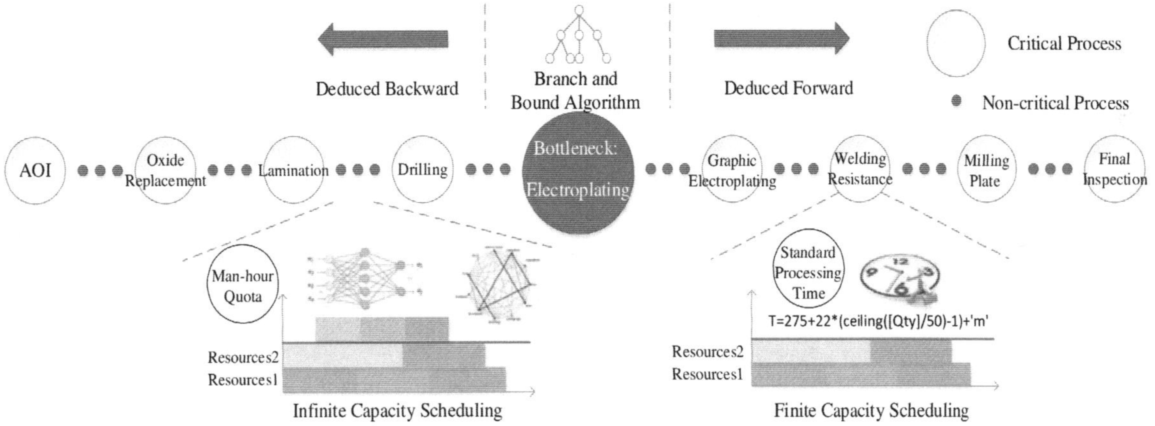

Figure 3. The Diagram of the Whole Production Scheduling in PCB Workshop

4. Experimental results

4.1 Scheduling optimization results of bottleneck process

Taking 100 orders as an example, branch and bound algorithm and genetic algorithm are used to optimize bottleneck process scheduling by the same encoding method, and the results are compared as shown in table 1.

Table 1. Experimental results of branch and bound comparison with genetic algorithm

Processing time	40, 100, 120, 140, 160	Branch and bound algorithm			Genetic algorithm		
		Parameter	Runtime	Makespan	Parameter	Runtime	Makespan
Number	Experiment 1 40, 25, 10, 15, 10	10	215.9	753.8	50	2725.4	754.8
		50	810	748.8	100	5745.4	752.4
	Experiment 2 40, 30, 10, 15, 5	10	205.5	734.6	50	2814.1	736.6
		50	655.5	732.8	100	5863.5	734.6
	Experiment 3 30, 30, 20, 15, 5	10	178.1	778.6	50	2723.1	785.2
		50	621.4	778	100	5818.5	783
	Experiment 4 50, 15, 20, 5, 10	10	263.3	719.6	50	2814.1	719
		50	838.8	715	100	5916.1	718.2
	Experiment 5 20, 25, 20, 25, 10	10	179	862.8	50	2704	869.6
		50	704.2	860.4	100	5619.5	868

The unit of runtime in the table is milliseconds. In branch and bound algorithm, the parameter represents the number of solutions generated randomly each time. Node cleaning is performed every 50000 times, and the first 100 items are retained each time. In genetic algorithm, the parameter represents the population size, and the reproductive algebra are 1000. The result of makespan is the average result of 10 runs under this parameter. From the results, it can be seen that the branch and bound algorithm is slightly better than the genetic algorithm for each case, and the runtime is much lower than the genetic algorithm.

4.2 Prediction and analysis results of man-hour quota for non-critical processes

There are four main parameters in the neural network model, learning factor, momentum factor, training times and allowable errors. By the methods of control variables and cross-parameter experiments, the accuracy is the highest when the training times are 1000, the allowable error is 1.00E-07, the learning factor is 0.1 and the momentum factor is 0.1.

Table 2. Cross-parameter experiment of learning factor and momentum factor

Process	Electronic testing	Milling plate	Graphic electroplating
Unrevised	64.0%	63.7%	63.0%
Revised	73.39%	78.61%	70.59%

Experiments on different processes with this parameter show that the performance is better and stable. Therefore, all experiments are carried out according to this set of parameters. The Apriori algorithm is used to obtain the association rules. Thus, the forecast results of man-hour quota can be modified according to the modified rules. After modifying, the accuracy of electronic testing prediction reached 73.39%, as show in table 2. In addition, this method is used to analyze the milling process and graphic electroplating, and the accuracy is 78.61% and 70.59% respectively. Therefore, this method is considered to be effective and feasible.

4.3 Results of overall production scheduling in PCB workshop

In this paper, the example of production scheduling experiment is carried out by taking the orders of one day in actual production. The results of production scheduling are shown in Figure 4.

(a) Gantt chart of electroplating process

(b) Gantt chart of graphics electroplating (c) Gantt chart of Lamination

Figure 4. Gantt charts of scheduling results

5. Conclusion

Current research studied the planning and scheduling method for a real PCB production system in Guangzhou China. A mixed integer programming model is built for electroplating process which is the bottleneck of the system. A branch and bound method is adopted to optimize its sequence. Furthermore, current paper proposed a method based on BP neural network and association rule analysis to predict the man-hour quota of non-critical processes. Experimental results indicate that the proposed scheme can optimize the sequence of bottleneck effectively and predict the man-hour quota accurately. Based on the optimal scheduling on bottleneck and accurate prediction of man-hour quota in non-critical processes, the suitable scheduling results of the whole PCB system can be figured out rapidly. In future, there are several directions needed to do more in-depth investigations. For example, it's necessary to improve prediction accuracy. Also it needs developing intelligent algorithms and making more experiments to improve the solution much better.

Acknowledgements

This work has been supported by MOST (the Ministry of Science & Technology of China) under the grants No.2012AA040909, 2012BAH08F04, & 2013AA040206, and by the National Natural Science Foundation of China (Grants No.51035001, 50825503, 71271156, & 51705379).

References

[1] Marcogaretti, Alessandropozzetti, Robertotavecchio Production scheduling in SMT electronic boards assembly[J]. Production Planning & Control, 1996, 7(2):197-204.

[2] Lim J M. A genetic algorithm for a single hoist scheduling in the printed-circuit-board

electroplating line[J]. Computers & Industrial Engineering, 1997, 33(3):789-792.
Manufacturing, 1998, 08(3n04):193-207.

[3] Kim Y D, Lim H G, Park M W. Search heuristics for a flowshop scheduling problem in a printed circuit board assembly process [J]. European Journal of Operational Research, 1996, 91(1):124-143.

[4] Dessouky M M, Leachman R C. Dynamic models of production with multiple operations and general processing times[J]. Journal of the Operational Research Society, 1997, 48(6):647-654.

[5] Kim Y D, Kim J U, Lim S K, et al. Due-date based scheduling and control policies in a multiproduct semiconductor wafer fabrication facility[J]. IEEE Transactions on Semiconductor Manufacturing, 1998, 11(1):155-164.

[6] Dabbas R M, Chen H N, Fowler J W, et al. A combined dispatching criteria approach to scheduling semiconductor manufacturing systems[J]. Computers & Industrial Engineering, 2001, 39(3-4):307-324.

IOP Conf. Series: Materials Science and Engineering **565** (2019) 012005 doi:10.1088/1757-899X/565/1/012005

Research on Automatic Flow-shop Planning Problem Based on Data Driven Modelling Simulation and Optimization

Weikang Fang, Zailin Guan, Dan Luo, Cong He, Hao Wang and Chuangjian Wang*

State Key Lab of Digital Manufacturing Equipment and Technology, School of Mechanical Science and Engineering, Huazhong University of Science and Technology, Wuhan 430074, China

*Email: wangchj@wust.edu.cn

Abstract. With the increasing demand for individualization, new smart manufacturing models are implemented in order to adapt to the market demands of high quality, multi-variety and fast delivery. However, new problems in production and operation are urgently needed for research. It is especially significant to study the automatic flow workshop planning based on data-driven modelling and simulation. Firstly, according to the rapid analysis of various data of the workshop, UML-based object-oriented data model is formulated. Then the application module of data-driven simulation modelling is established by using the simulation platform. The bottleneck recognition based simulation optimization module for the tabu search algorithm with guiding rules is designed. Data-driven modelling makes the model more flexible and reusable which can run a variety of different experiments by modifying the external data to drive simulation model. Finally, experiments are conducted to verify the effectiveness of the simulation optimization method proposed in current study. In addition an optimized production line configuration plan is provided.

1. Introduction

With the aggravation of the industry competition, how to quickly seize the growing market becomes a critical task for enterprises. Through the upgrade of the current production mode, many companies adopt advanced intelligent hardware and software systems to implement a new mode of intelligent manufacturing. Thereby increasing production capacity and product quality, enhancing the company's own competition, expanding the market and reducing production costs [1]. For the planning of intelligent manufacturing lines, simulation technology could handle complex systems that cannot be processed using mathematical models, accurately describe real-world conditions and identify key factors that affect the system. It performs simulation and solution verification quickly and gives a variety of resource allocation schemes and compares various indicators. Modelling is the premise and basis for conducting simulation experiments, which is also one of the most complicated stages in the implementation of simulation projects. However, traditional simulation technology consumes a lot of manpower and time costs, and cannot respond quickly to changes in the production floor [2]. There is an urgent need for a fast, real-time simulation system to accurately evaluate and optimize the production system.

However, the construction cost of smart manufacturing workshops is very high, because the imbalance of resource allocation leads to high cost of redesign and construction. At the beginning of the design, it is necessary to not only express the physical structure of the system, but also describe the operational logic of the production system. Therefore, the application of data-driven modelling and simulation technology is proposed.

Data is the driving force for modelling, directly serving the development, operation, and evaluation of automated modelling and simulation. Reconstruction of the production line depends on the data changes. Data-driven modelling allows modelers to provide modifications and corresponding additions at the input data level, which can lead to rapid construction of simulation models.

Generally, the simulation system and various simulation softwares have dedicated formats to provide data import and export interfaces. It takes a lot of time to convert according to the interface. Therefore, it is necessary to establish a unified data model storage. The research has proposed different CMSD frameworks to solve the problem that different simulation softwares can't be converted according to different formats [3]. Lee [4] builds data transfer mechanism through XML and ODBC to realize shop floor data to simulation model Data conversion. Fournier J et al [5] divided the data transformation into two steps. First map to the XML data model, and then convert to simulation code through XML and XSLT data model to realize the interactive integration of MRP, ERP and other systems with simulation models.

James Tannock et al [6] used data-driven simulation to support the design and improvement of supply chains. Based on the physical layout data of assembly line, Wang Junfeng [7] implemented the data-driven generation simulation model for the automobile assembly line using the Arena software. Wang Guoxin et al [8] described the system structure hierarchically and establishes corresponding control models respectively. Dynamically complete simulation, experimentation, verification and optimization based on production data, experimental parameters, control parameters, statistical data. Meng et al [9] developed a data-driven modelling and simulation framework for MHS of coal mines to automatically generate a discrete-event simulation model based on current MHS structural and operational data.

Data-driven modelling and simulation technology is a system integration application. The integration of simulation with other information systems helps advance the study of system modelling and simulation. Modelling and simulation optimization is an organic whole. While considering the versatility and reusability of the model, the robustness, the parameterization of the optimization algorithm, and the customization characteristics are also needed to consider.

Aiming at the rapid planning and design of intelligent manufacturing workshop, the current research proposed a data-driven modelling and simulation and optimization planning method to study the practical application of data-driven simulation modelling technology in the field of intelligent manufacturing. Focusing on the execution layer of the intelligent workshop, analyze the characteristics of the line and building a data model in the database including workshop layout data and simulation running logic data, and design a data-driven simulation modelling module on the simulation software platform. The fast and accurate generation, reusability, maintainability and other performance of the data-driven simulation model enhances the rapid analysis of the manufacturing floor. Finally, the simulation optimization based on the tabu search algorithm in the simulation model is designed to optimize the automated production line configuration. Through the simulation model design optimization program, the planning and design efficiency of the on-site workshop is improved, and the intelligent level of the manufacturing enterprise is continuously promoted. Production line planning framework based on data-driven modelling and simulation as shown in Figure 1.

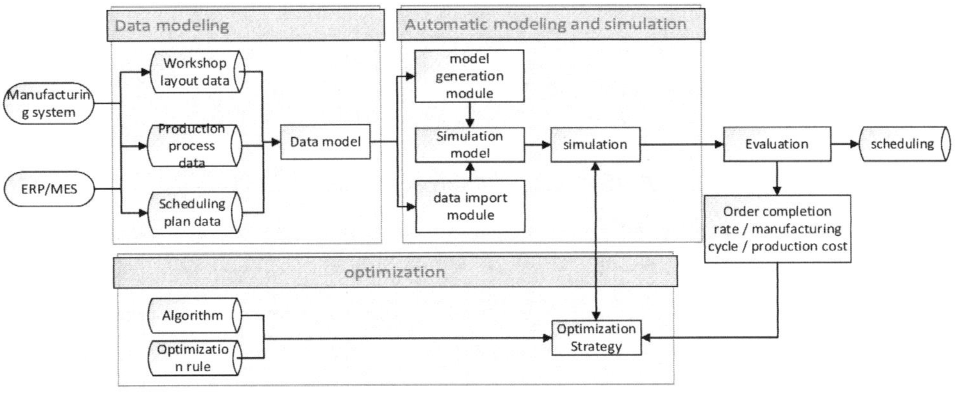

Figure 1. Production line planning framework based on data-driven modelling and simulation

2. Data-driven modelling and simulation for the automated production line workshop

Current research formulated a data model in the database after analyzing the characteristics of the data through object-oriented analysis, and then analyzed the simulation modelling logic and running logic of the automated production line workshop. The layered modelling method is used to build the customized entity in the simulation platform. And the physical layout of the workshop is automatically generated, running through the logical data driven model.

2.1 Data model of the workshop

The production line workshop model consists of two parts: the production line and the logistics system. The production line mainly completes the material processing process, and the logistics system is responsible for the entire transportation process of the materials in the entire workshop. The robot is responsible for the refueling action of the auxiliary machine, and the different number of machines and robots constitute the smallest production unit of the production line. The incoming and finished products of the production line are stored in the line buffer of each assembly line to form a complete production line module.

After sorting out the basic components and processes of the workshop, the workshop is modeled from the perspective of data modelling. Use the class diagram in UML to represent the elements of the workshop, which describes the abstract classes in the workshop and the relationship between classes and classes. The data model of the workshop is constructed through the relational data table of the database management system.

2.2 Data driven modelling and simulation of the workshop

At the beginning of the simulation, modeling must be based on the entity of the workshop. Then a complete workshop model should be established after the analysis of the logical flow of the complete and accurate workshop elements.

Using the layered modelling method, the resource model is transformed into the simulation input data from the Access data interaction platform to establish the simulation model in the automatic pipeline workshop. The data-driven simulation model generation module is developed in the simulation software platform, and the product order data in the data model is imported for dynamic simulation to drive the simulation model to build and run.

There are three modules in automatic modelling and simulation: model generation module, data import module, and operation control module. The model generation module is shown in Figure 2. The model generation module automatically generates the static simulation model of the workshop by reading the physical layout data in the data-model, automatically generating the necessary elements such as the intelligent entity, elements, data tables and simulation logic of the simulation model. Data import module: The data table in the simulation model is associated with the data table of the database. The module calls the data import module before the simulation runs, and the data in the database is imported into the simulation model to drive the simulation model to run.

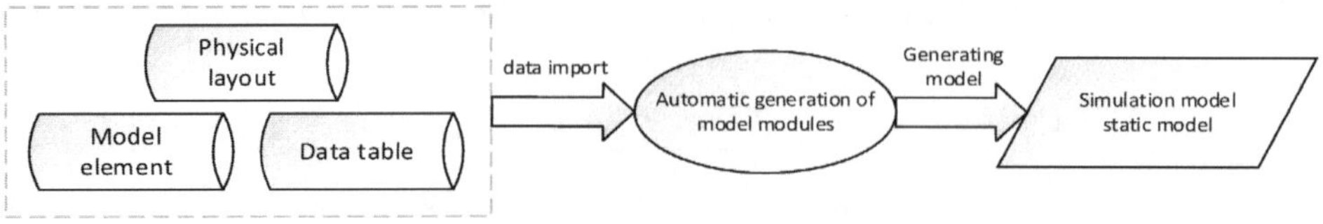

Figure 2. Automatic generation of static models

The essential to data-driven modelling is to accurately map the data in the data-model. Design a data-driven simulation model generation module developed by API, and drives the simulation model to build and run by importing data from the data-model, as shown in Figure 3.

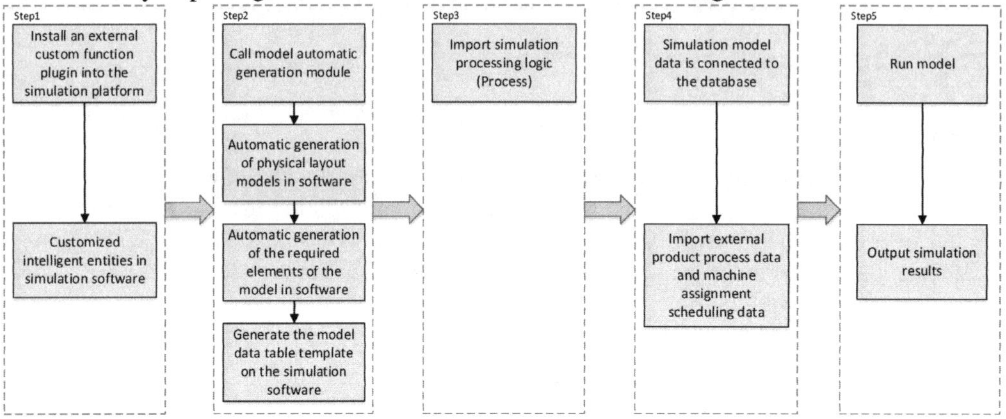

Figure 3. Data driven simulation modelling and simulation

The simulation needs to be defined separately operating logic (Process) to implement complex logic simulation model. In the data-driven modelling, the XML file which saving the pre-defined simulation logic is imported into the model. Then the Process processing flow is automatically generated in the model, so that the simulation model is organized in accordance with predetermined rules.

3. Data-driven simulation optimization of machine allocation planning

Based on the data-driven simulation modelling, the optimization algorithm library is added to optimize the simulation model for the unreasonable allocation of resources. The results of the simulation experiment are evaluated by the specified indicators. The optimization program feeds back the output of the simulation and adjusts the configuration parameter data of the simulation model if optimization is needed. By repeatedly performing the above process, the parameters of the model are continuously optimized. Finally simulation gives the scheduling scheme to be executed.

In the planning and design, it is easy to ignore the influence of the manipulator on the production line process resulting in the unreasonable configuration of the assembly line, machine and robot, which makes it impossible to identify the real bottleneck on the assembly line. Aiming at the problem of unreasonable allocation of production line resources, the current research designed an algorithm simulation optimization module with guiding rules based on bottleneck recognition which obtains the optimization plan of production line configuration.

The process that defines the maximum load on the machine is the bottleneck process [10]. The bottleneck identification method in the simulation model with process load and robot load as indicators is used to determine the bottleneck position on the production line.

The basic parameters are shown as following Table 1.

Table 1. Model parameters

L	Distribution of workstations on the production line	t_n	The time taken by the robot to transfer the tray at the end of the production line
Q	Number of processes	t_c	Material cleaning time in the process
S	Number of workstations	r_j	robot in the production line
N	Number of robots	m	machine in the production line
P_i	Sequence of operations, "i" indicates the number of iterations	a_i	Number of materials replaced by r_i during simulation
R_j	Robot sequence, "j" indicates the number of iterations	t_x	Total robot movement time during simulation
P_0	Initial value of the sequence of operations	p_i	Process in the production line
R_0	Initial value of the robot sequence	ω	Work load rate of machine m_i
φ	The random transformation number of the solution in the algorithm	n_i	Number of machines included in the process p_k
TP_i	Output of the production line at the ith iteration	T_i	Total output of machine m_i during simulation
t_p	Processing time in the process	ω	Work load rate of robot r_i
t_r	The time taken by the robot to replace the material in the process	ω	Resource load rate of process p_k

The calculation for load factor of processing machine and robot are given in Equations (1) and (2). By calculating the workload of each machine, the average load of all machines in the process is taken in Equation (3) which determine the position of the bottleneck process.

$$\omega_{m_i} = TP_{m_i} \times \left(t_p + t_r + t_c + t_m/2\right)/t_{run} \tag{1}$$

$$\omega_{r_j} = a_{r_i} \times (t_r + t_x + t_m/2)/t_{run} \tag{2}$$

$$\omega_{p_k} = \sum_{i}^{n_{p_k}} \omega_{m_i}/n_{p_k} \tag{3}$$

According to the workload, the process and the robot are sorted. The process with the largest workload is known as the bottleneck process. At the same time, the robot with the largest workload will also affect the production capacity of the production line. Therefore, the guiding rules based on the workload rate of the robot and the process could adjust the machine allocation plan including the assignment of processes, robots, and workstations.

The simulation input is controlled by the external program. And the simulation model is continuously called for calculation to optimize the production line configuration. When adjusting input parameters, set up a rule bootstrap to quickly find an optimized configuration of the production line. The tabu search algorithm optimization process is shown in Figure 4. Through the work load of the production line process and the robot, the machine distribution plan is adjusted to obtain a better distribution plan of the production line machine.

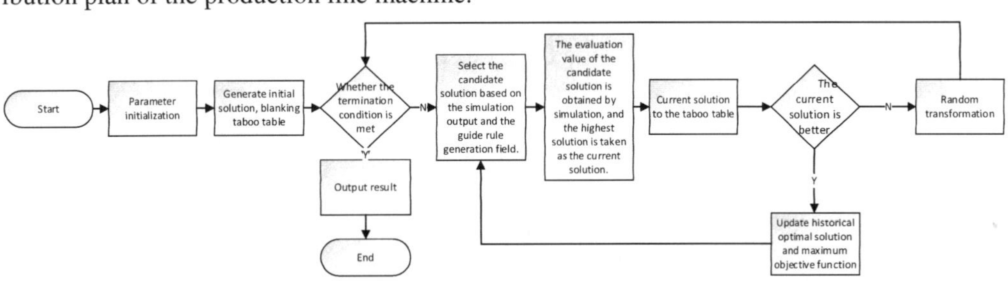

Figure 4. Optimization algorithm flowchart

4. Experimental result

It is considered that model verification and validation is extremely important in simulation practice. The data-driven simulation modelling is verified through simulation experiments. The simulation logic and output comparison of the model are verified to test the feasibility of the data-driven modelling technology and the effectiveness of the simulation model.

Experiment 1: Design a simplified assembly line workshop with 2 production lines, 6 robots, 1 AGV trolley and 2 loading points. The workshop processes both products P1 and P2. The work of P1 is j1, j2, j3, and the work of P2 is j4, j5, j6, which are processed on the production line LineA and the production line LineB respectively.

Run the simulation model for 48 hours, and the output and utilization of each machine and the robot utilization are imported into the Result during operation. The simulation output for comparison with the theoretical output calculated according to the formula.

Table 2. Experimental output versus theoretical output

Process id	Product	Work	Experimental output	Theoretical output	Error(%)
pro1	P1	j1	1247	2362	-
pro2	P1	j2	1290	2310	-
pro3	P1	j3	1328	1470	-9.66
pro4	P2	j4	922	934	-
pro5	P2	j5	911	1020	-
pro6	P2	j6	901	932	-3.33

As shown in the Table 2, the **yellow marked data** in the table is the process with the lowest processing capacity meaning the production line bottleneck which need be compared when comparing the errors. The comparison error between the simulated output and the theoretical output of the bottleneck process are -9.66% and -3.23% of pro3 and pro6 within acceptable limits. The overall output of each process in a pipeline is roughly equal due to the JIT production in the simulation logic. The validity and correctness of data-driven modelling is verified.

Table 3. Robot utilization

Robot id	r1	r2	r3	r4	r5	r6
Utilization (%)	65.36	64.41	66.35	50.63	28.79	41.17

The theoretical output is larger than the simulation output because the theoretical output calculation does not take into account the situation where the robot services multiple machines. As shown in Table 3, the maximum utilization rate of the robot reached 66.35%. The influence of the manipulator on the output of the processing machine cannot be neglected. Therefore, an emergency search algorithm flow with guiding rules is designed to adjust the machine allocation plan. Finally a better production line machine distribution plan is obtained.

Experiment 2: 7 kinds of production lines with different configurations, respectively processing 2 kinds of products. The initial data is as shown in Table 4.

Table 4. Initial data

Line id	L	Q	S	N	t_p	t_p	t_p	t_p
LineA	[2,2,2,2,2,2,2,2]	3	8	8	[180,300,400]	[30,30,30]	[25,25,25]	[0,0,15]
LineB	[2,2,2,2,2,2,2,2,2,2,2]	3	11	11	[180,300,400]	[30,30,30]	[25,25,25]	[0,0,15]
LineC	[1,2,2,2,2,2,1,2,2,1,2]	3	11	4	[180,300,400]	[30,30,30]	[25,25,25]	[0,0,15]

LineD	[2,2,2,2,2,2,2,2,2,2,2]	3	11	5	[180,300,400]	[30,30,30]	[25,25,25]	[0,0,15]
LineE	[2,2,2,2,2,2,2,2,2]	3	9	9	[240,480,360]	[30,30,30]	[25,25,25]	[0,0,15]
LineF	[1,2,2,1,2,2,1,2,2]	3	9	3	[240,480,360]	[30,30,30]	[25,25,25]	[0,0,15]
LineG	[2,2,2,2,2,2,2,2,2]	3	9	4	[240,480,360]	[30,30,30]	[25,25,25]	[0,0,15]

Run the model by importing the production process and the machine-assigned data. Then the tabu search algorithm is optimized for production line configuration. Initial solution and optimization solution are shown in Table 5.

Table 5. Initial solution and optimization solution

Line id	P_0	R_0	φ	TP_0	P_i	R_i	TP_i
LineA	[2,3,3]	[1,1,1,1,1,1,1,1]	2	1153	[2,3,3]	[1,1,1,1,1,1,1,1]	1153
LineB	[3,3,5]	[1,1,1,1,1,1,1,1,1,1,1]	3	1495	[2,4,5]	[1,1,1,1,1,1,1,1,1,1,1]	1154
LineC	[2,4,5]	[3,3,3,2]	13	886	[3,4,4]	[2,3,3,3]	1275
LineD	[2,4,5]	[2,2,2,2,3]	19	1298	[3,4,4]	[1,3,3,2,2]	1515
LineE	[2,4,3]	[1,1,1,1,1,1,1,1,1]	2	973	[2,3,4]	[1,1,1,1,1,1,1,1,1]	1205
LineF	[2,3,4]	[3,3,3]	4	751	[2,4,3]	[2,4,3]	886
LineG	[2,3,4]	[2,2,2,3]	7	960	[2,4,3]	[2,2,3,2]	1174

Table 6. Optimized production lines capacity

Line id	LineA	LineB	LineC	LineD	LineE	LineF	LineG
Increase (%)	0	3.3	43.9	16.7	23.8	18	22.3

The production line optimization scheme is obtained through the optimization of the simulation algorithm. As shown in Table 6, the results of the simulation optimization are compared. The average productivity through algorithm optimization has increased by 18.3%, and the production line has increased by up to 43.9%. The results show that the algorithm can improve the production capacity of the production line by adjusting the machine allocation plan under the existing pipeline configuration.

5. Conclusion

In current research, a data-driven modelling and simulation method is proposed to effectively simulate the intelligent workshop simulation system. Data-driven modelling technology integrates data with other systems. The technology automatically generates and runs real-time data-driven simulation models which could responds to requirements to assist the site in making more accurate decisions. When the planning and design scheduling does not achieve the predetermined aim, different simulation experiments can be run by modifying the external database data-driven model, which can not only promote the accuracy of modelling and reduce the complexity of re-modelling, but also enhance the flexibility of the simulation model. Generality and portability improve planning efficiency. The combination of modelling techniques and optimization methods extend the range of applications for simulation optimization. It is not only effective for the intelligent workshop production line design, but also valuable for production lines with diverse order requirements, similar production line structure, scale and occasional differences.

Acknowledgements

This work has been supported by MOST (the Ministry of Science & Technology of China) under the grants No.2,12AA,4,9,9, 2,12BAH,8F,4, & 2,13AA,4,2,6, and by the National Natural Science Foundation of China (Grants No.51,35,,1, 5,8255,3, 71271156, &517,5379).

References

[1] Guan, Z.L., et al. TOC/DBR based production planning and control in a manufacturing system with multiple system bottlenecks. in IEEE International Conference on Industrial Engineering & Engineering Management. 2007: IEEE.

[2] Wang, N., S.Q. Li and J.F. Wang, A Data Driven Modeling and Simulation Methodology for Automotive Assembly Plant. Advanced Materials Research, 2011. 346: p. 228-235.

[3] Qiao, G.X., F. Riddick and C. Mclean. Data driven design and simulation system based on XML. in Proceedings of the 2003 Winter Simulation Conference. 2003.

[4] Lee, Y.T. and L. Yan. Data exchange for machine shop simulation. in Proceedings of the 2005 Winter Simulation Conference. 2005.

[5] Fournier, J. Model building with core manufacturing simulation data. in Proceedings of the Winter Simulation Conference. 2011.

[6] Tannock, J., et al., Data-driven simulation of the supply-chain—Insights from the aerospace sector. International Journal of Production Economics, 2007. 110(1-2): p. 70-84.

[7] Wang, J., et al., Data driven production modeling and simulation of complex automobile general assembly plant. Computers in Industry, 2011. 62(7): p. 765-775.

[8]. Yan, Y. and G. Wang. A job shop scheduling approach based on simulation optimization. in IEEE International Conference on Industrial Engineering & Engineering Management. 2007: IEEE.

[9]. Meng, C., et al., Data-driven modeling and simulation framework for material handling systems in coal mines. Computers & Industrial Engineering, 2013. 64(3): p. 766-779.

[10]. He, C., et al., Set-partitioning-based heuristic for balancing and configuration of automated flexible machining line. International Journal of Production Research, 2018. 56(9): p. 3152-3172.

Design of Secure Enhanced Privacy Protection Electronic Hotel-card Based on QR Code

Eryang Chen, Ansheng Ye*, Fang Miao, Wenjie Fan and Yi Jiang

School of Information Science and Engineering, Chengdu University, Sichuan province, China

*Email: cey@cdu.edu.cn

Abstract. The development of the Internet and economy urges researchers to look for a safe and more economical way to secure online hotel-users' private information. This study is rooted in online-hotel user personal privacy and some information protection. The traditional RFID hotel-room card was replaced by the proposed QR e-ID, which combines cryptography, verification code, passwords and multi-directional identity authentication module. All of the privacy information should be registered in the data register center (DRC) with encrypted mode. In order to ensure safety in case if the QR e-ID is used by unknown person it needs a password. The main goal of this study is to investigate how the co-use of computers and mobile devices protect online hotel-users' private information, in addition, reduce the hotel operating costs effectively. The results of the study also provide guidelines to IT developers for future directions in virtual assets protection.

Key words: Personal Privacy, QR e-ID, Multi-directional Identity Authentication, Virtual Assets Protection

1. Introduction

With the development of the Internet, 4G/5G technology and economy, for one thing, smartphone has become an impartible part of life for many people, for another, more and more people choose to visit other places during holidays, booking hotels online. It provides utility by assisting front office process before checking in, however, it has a underlying risk because personal privacy and some information may be exposured in public [1]. Because the information privacy and security issues have not attracted sufficient attention from individual uses, or hotel backstage database may be attacked by hackers [2]. Although personal privacy and some information protection has been discussed on media for quite a long time [3].

For example, credit reporting firm Equifax (US) suffered one of the worst security breaches in history when it announced that sensitive data--including Social Security numbers and driver's license numbers of more than 147 million consumers were exposed to hackers from mid-May to July 2018 [4]. Both The New York Times and Observer broke the news that 50 million profiles of Facebook users were "harvested" without their consent to a consulting firm, Cambridge Analytica in March 2018 [5]. In August 28th 2018, the Internet broke the news that the users data of HuaHua Groups chain hotel were suspected to have leaked. About 500 million information leaks cover all personal information, and the data include Hanting Hotel, Mercure, Xiyu, Wanxin, Novotel, Mercure, CitiGo, Orange, Season, Star Trek, Ibis, Elite, Haiyou and other hotels [6].

To provide an alternative of authentication for online accounts, QR code (Quick Response Codes) is introduced. QR codes were initially used in Marketing for advertising purpose for its strong ability to store identification information of the advertiser and its convenience to use [7]. For example, the only way to log in WeChat in PC is to scan the QR code by a mobile phone or a pad. This study would help practitioners understand how QR codes mitigate online users' fears of threats to their privacy and security.

2. Literature review

QR code was initially invented for the automotive industry in Japan in the early 1990s. Similar to bar codes, QR codes technically are machine-readable optical labels using two-dimensional barcode to store information of the item to which it is attached or associated. A QR code uses four standardized encoding modes such as numeric, alphanumeric, byte/binary, and kanji (the subset of Chinese characters used in the Japanese language) to store data efficiently [8].Graphically, a QR code consists of black squares arranged in a square grid on a white background, including five areas: finder pattern, alignment pattern, timing pattern, quiet zone, and data area, and each area has its unique functionality [9].

The QR code became popular outside the automotive industry due to its fast readability and greater storage capacity compared to standard UPC barcodes. The symbol versions of the QR Code range from Version 1 to Version 40 [10]. Each version has a different module configuration, which refers to the number of modules contained in a symbol, commencing with Version 1 (21 × 21 modules) up to Version 40 (177 × 177 modules) Each QR Code symbol version has the maximum data capacity, according to the amount of data, character type and error correction level.

Today, marketers are creating exciting, new cross media strategies by including digital QR codes on printed pieces across the marketing mix; from collateral to bill statements to signage and product placement, the applications are endless. In other words, QR Codes are called the link between the Electronic and Paper Worlds.

G.PRABAKARAN [11] proposed the authentication system based privacy protection QR-code, that allows the user to safely enter credentials and information to transfer money after launching LIVE-CD on stand alone in place and avoiding the possibility of entering credit card details (send or receive money without using of credit cards). The entered information is encrypted by common key crypto system and stored in a QR (quick response) code. YAN Wen-bo [12] proposed a privacy-preserving scheme for logistics systems based on 2D code and information hiding.

3. Research models

3.1. DRC

Whether information security or network space security, data is the most core security goal. Especially in the open environment, the safety of data and the safe application of data face great challenges. The data is encrypted and stored in the data registration center (DRC), and the system can only be used legally after authorization. This approach makes sense for data security and data security applications [13].

3.2. QR e-ID

A QR e-ID will be automatically generated for the current user after the order is completed. The QR e-ID contains the basic information of the guest room and the identity information of the guest. Therefore, the security of the QR e-ID is an important issue. The DRC plays an important role in improving the QR e-ID safety. Figure 1 shows Mutual information between the database and the QR e-ID.

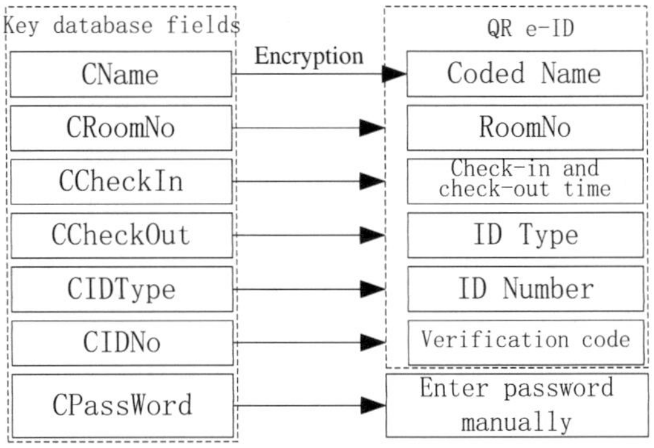

Figure 1. Mutual information between the database and the QR e-ID

In order to improve the security of the QR e-ID as much as possible, the security measures are as follows:

(1) The security of background data is ensured by the DRC.

(2) The customer's name is coded as a combination of letters. If the tenant's name is in Chinese, the initials of the Chinese phonetic alphabet will be extracted; If the tenant's name is in English, the initial combination of English letters will be extracted. So that, even if the QR e-code was stolen,it is difficult to obtain the customer' real information directly.

(3) A set of verification codes is coded according to the guest and booking information. Even if the basic information of the customer and reservation is compromised, the correct verification code cannot be generated, which can avoid the forgery of the QR e-code.

(4) Each QR e-code requires the user to set up a set of personal passwords, whose length is no less than 6.

3.3. Illegal user identification

According to relevant national laws and regulations, guests must provide their ID card to verify their real information when they check-in for the first time. IT combines cryptography, verification code, and passwords. And this is called the multi-directional identity authentication module shown in Figure 2.

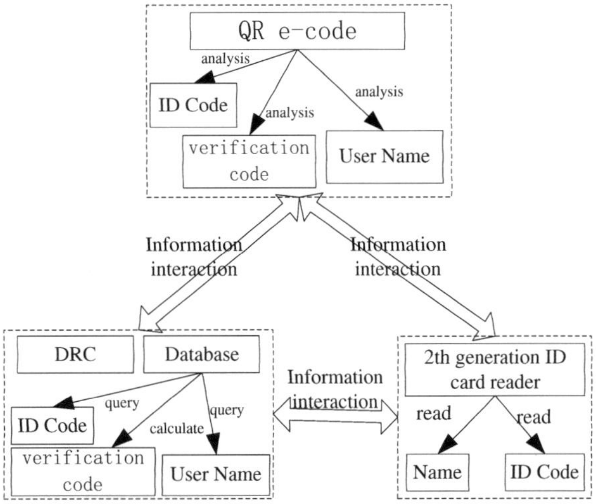

Figure 2. The multi-directional identity authentication module

The name and the ID code read by the 2th generation ID card reader is marked as Name_A and ID code_A; the name, the ID code and the verification code read from the QR code is marked as Name_B,

ID code_B and verification code_B; With ID code_A as the query keyword, DRC and database are queried. The results are marked as Name_C and ID code_C, and verification code_C is calculated.

When customers check in, the verification process is as the follows:

(1) Only when Name_C or ID code_C is not NULL, then real name authentication passed.

(2) Only then Name_B = Name_C, ID code_B = ID code_C and verification code_B = verification code_C, then the QR e-code is not a fake.

(3) When the password entered by the user matches the system successfully, the guest room will open automatically.

(4)A set of passwords (the password length is not less than 6) is set by the user. After being encrypted by MD5, the password exists in the database as ciphertext, and the tenant enters the password to verify the system.

3.4. Calculation mode of the verification code

The ID number in mainland China is 18 in length, and the bit 18 is a number or a letter "X". Bit 4 to bit 10 is hided when the QR is coded. And bit 11 to bit 18, check-in time, check-out time and the length of passwords combines to a set of number, according to which the verification code is coded. From beginning to end, each two-digit number is coded as a letter. And the mapping rules f(m)are as follows:

$$\begin{cases} 01 -> \ 'A' \\ 02 -> \ 'B' \\ \quad \ldots \\ 26 -> \ 'Z' \end{cases} \tag{1}$$

Take the two-digit number n1n2 as an example, the code result α can be expressed as:

$$\alpha = \begin{cases} Null, & n_1n_2 = 00; \\ f(n_1n_2), & 01 \le n_1n_2 \le 26; \\ f(n_1)f(n_2), & n_1n_2 \ge 26; \end{cases} \tag{2}$$

4. Experiment

4.1. Example

This section shows the actual effects of a QR e-id card, illustrated with the data of a virtual customer. Customer information and room reservation information are as follows:

Name: MaJianlong; RoomNo:303; Check-in Time: May 1, 2018; Check-out Time: May 3, 2018; Certificate type: ID Card; Certificate No:510123199606270066. The user's personal password is 0123456abc. Based on the above information system, the verification code can be calculated as :FBGBHFFTREATRECJ.

The QR e-card is shown in Figure 3.

Figure 3. The QR e-card

The information contained in QR is: CName:MJL; CRoomNo:303; CCheckIn:20180501; CCheckOut:20180503; CIDType:001; CIDNo:510*******06272866; Parameter_VerificationCode: FBGBHFFTREATRECJ.

During the valid time of booking, the customer can check in after entering the personal password "0123456abc" as prompted by the system.

4.2. Security evaluation

The evaluation of the security is shown in Figure 4.

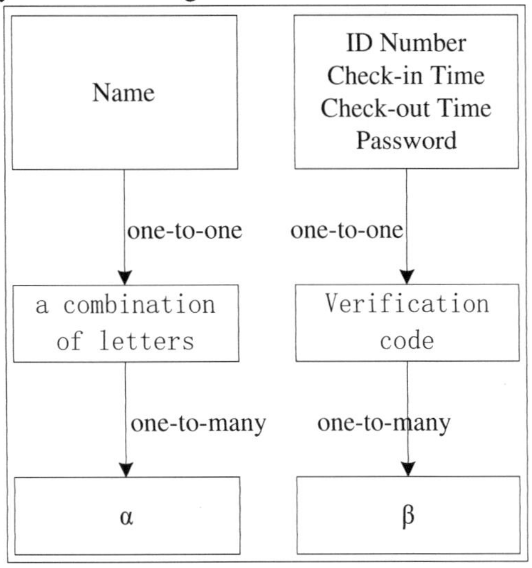

Figure 4. The evaluation of the security

The security of the system can be illustrated by the example in section 4.1.

According to the homophone characteristics of Chinese characters, α. Bit 11 to bit 18 of the ID Number, check-in time, check-out time and the length of passwords combines to a set of number '06272866201805012018050310'. And there are at least 2 scenarios for β. In addition, the password whose length is 10 consists of capital letters, small letters, and Numbers. Therefore, the number of passwords can be expressed as (3):

$$f(\gamma) = (26+26+10)^{10} \approx 3.0331\text{e+}17 \tag{3}$$

Based on the above description, there are more than 3*2*f(γ) possibilities for one QR e-code card. The system has high security.

5. Conclusion & future enhancement

This paper presents an alternative method for authentication in hotel room management. Basically this method involves the Check-in process by keeping the QR-CODE at the core part. Here we have made a progress in the user perspective of implementation of the technology. Use of the traditional RFID hotel-room card by using the QR e-code card, which co-use computers and mobile devices to protect online hotel-users' private information, in addition, reduce the hotel operating costs effectively.

The whole process is carried out by encrypting the data of hotel-users' private information and generating the QR e-code. This project can be further improved by increasing its application areas by implementing it in virtual assets protection, in addition to the electronic hotel-card.

Acknowledgments

The work is supported by National Key Research and Development Program under Grant 2016YFB0800600, Opening Fund of Geomathematics Key Laboratory of Sichuan Province (csxdz201710), Key Laboratory of Pattern Recognition and Intelligent Information Processing, Institutions of Higher Education of Sichuan Province, Chengdu University, Funding MSSB-2018-0. Sichuan support Plan(No.2016FZ0112).

IOP Conf. Series: Materials Science and Engineering **565** (2019) 012006 doi:10.1088/1757-899X/565/1/012006

References

[1] Crossler, R.E., et al., Understanding Compliance with Bring Your Own Device Policies Utilizing Protection Motivation Theory: Bridging the Intention-Behavior Gap, in Journal of Information Systems. 2014, American Accounting Association. p. 209-226.
[2] Meso, P., Y. Ding, and S. Xu. Applying Protection Motivation Theory to Information Security Training for College Students. Journal of Information Privacy & Security, 2013. 9(1): p. 47-67.
[3] Tank A H, Unde M M, Patel B J, et al. Storage and transmission of information using grey level QR (quick-response) code structure[C]// Advances in Signal Processing. IEEE, 2016:402-405.
[4] Credit firm Equifax says 143m Americans' social security numbers exposed in hack, https://www.theguardian.com/us-news/2017/sep/07/equifax-credit-breach-hack-social-security.
[5] Revealed: 50 million Facebook profiles harvested for Cambridge Analytica in major data breach, https://www.theguardian.com/news/2018/mar/17/cambridge-analytica-facebook-influence-us-election
[7] Wang J S, Zhang J N. Design and Realization of Hotel Management System[J]. Computer Engineering & Design, 2011, 2(1):1173-1176.
[6] Hotel group probes private info leak of 130 million clients, http://www.globaltimes.cn/content/1117459.shtml.
[8] Kroenke, D., Experiencing MIS, in 5/E. 2014, Prentice Hall. p. 696.
[9] Kan, T.-W., C.-H. Teng, and M.Y. Chen, QR Code Based Augmented Reality Applications, in Handbook of Augmented Reality, B. Furht, Editor. 2011, Springer: New York. p. 339-354.
[10] QR Code Tutorial, http://www.thonky.com/qr-code-tutorial/
[11] Prabakaran G,Bhakkiyalakshmi R, "Transmission of Data Using Arm Based Privacy Protection QR-code", International Journal of Engineering Development and Research (IJEDR), ISSN:2321-9939, Vol.2, Issue 2, pp.1458-1461, June 2014.
[12] YAN Wen-bo, YAO Yuan-zhi, et al.Privacy-preserving scheme for logistics systems based on 2D code and information hiding[J].Chinese Journal of Network and Information Security,2017,3(11):22-28.
[13] MIAO Fang. Data Oriented Security Architecture[J]. ZTE Technology Journal, 2016(1):19-22.

IOP Conf. Series: Materials Science and Engineering **565** (2019) 012007 doi:10.1088/1757-899X/565/1/012007

Generating Calligraphy Works Based on DiscoGAN

Yuqing Zhao[1*] and Guangyuan Fu[1]

[1]Xi'an research institute of high-tech, Shaanxi,710025, China

[*]Email: zoe_rabbi@126.com

Abstract. Copying the works of a good calligrapher is a necessary step for many beginners. However, due to the large number of Chinese characters, it is a challenge to automatically generate Chinese characters in the style of calligraphers. With the development of deep learning, the recognition of Chinese characters, style migration and other research have achieved remarkable results. However, there are few studies on the generation of calligraphy works, especially on the lack of paired training data. In this work, we propose a method for generating calligraphy works. This method is generally applicable to the generation of calligraphy works of various styles. And the validity of our method is verified by experiments on data sets.

1. Introduction

Calligraphy is a unique form of artistic expression of the beauty of Chinese characters in China and surrounding countries and regions deeply influenced by Chinese culture. Among them, Chinese character calligraphy is an original expressive art of the Han nationality, which is known as: wordless poetry, invisible dance, pictures without pictures, silent music. Like other literary theories, calligraphy theory includes both the technical theory of calligraphy itself and its aesthetic theory. From ancient times to the present, a group of outstanding calligraphers have emerged, among them Xizhi Wang (303-361,Eastern Jin Dynasty), Xun Ouyang (557-641,Tang Dynasty), Zhenqing Yan (709-784,Tang Dynasty), Gongquan Liu (778-865, Tang Dynasty), Mi Fu (1051-1107,Northern Song Dynasty) and so on have left a famous masterpiece. Long time ago, many famous works lost. The existing part also cannot cover all the commonly used words. As a result, more and more people are beginning to try to imitate more famous masterpieces in modern ways.

In this work, we formulate the problem of generating calligraphy works as a problem of learning the mapping from unpaired printed fonts to calligraphers' fonts. Our main contributions are:

- Based on DiscoGAN [1], the method of generating calligraphy works is proposed.
- Identity consistency loss is proposed to adapt to the specific application environment.
- The effectiveness and versatility of the proposed method are proved by experiments.

2. Related work

2.1 Reorganizing Chinese characters by decomposing strokes

Many of the previous ideas [2-4] were expressed as the basis of Chinese characters through the level of simple strokes. Decompose existing Chinese characters into stroke trees and establish an accurate correspondence between strokes. Combine new Chinese characters by re-typing. However, this method is difficult to guide the synthesis process with high regular statistics.

2.2 Neural style transfer

A major advance in deep learning is the neural style transfer, which was proposed by Leon Gatys et al in 2015 [5]. Style transfer refers to applying the style of the reference image to the target image while retaining the content of the target image. The key concept of implementing style transfer is consistent with the core idea of the deep learning algorithm: define a loss function to specify the desired goal and then minimize the loss.

$$\begin{aligned}\text{loss} &= distance(style(reference\_image) - style(generated\_image)) \\ &+ distance(content(original\_image) - content(generated\_image))\end{aligned} \quad (1)$$

Where distance is a norm function; Content is a function that takes an image and computes its representation; Style is a function that inputs an image and computes the representation of its style. Minimizing this loss will make the style of the generated image close to the reference image, and the content of the generated image also close to the target image, so as to achieve the style transfer. Although style transfer is applied to many images, the generation of calligraphy works involved in this paper is not entirely a matter of neural style transfer. In particular, the particularity of Chinese characters, such as: stroke position and Angle slightly change will lead to the change of different characters.

2.3 Generative adversarial networks (GANs)

Generative adversarial networks (GANs) [6] consists of a generator G and a discriminator D, which compete in a two-player minimax game: the generator attempts to generate realistic images to deceive the discriminator, while the discriminator attempts to distinguish the synthetic images from the real data. Formally, D and G are two players of min-max problem to find Nash equilibrium:

$$\min_G \max_D \mathbb{E}_{x\sim p_{data}(x)}[\log D(x)] + \mathbb{E}_{z\sim p_z(z)}\left[\log\left(1 - D\left(G(z)\right)\right)\right] \quad (2)$$

Goodfellow et al. (2014) proved that when $p_g = p_{data}$, this minimax game has the global optimal value, and under mild conditions (for example, G and D have sufficient capacity), p_g converges to p_{data}.

2.4 Matching image conversion

Pix2Pix [7] adopts the idea of cGAN [8] and takes the input image as the condition to Generative adversarial networks. In the design of the network structure, Pix2Pix basically refers to the structure of DCGAN [9], using convolution layer, batch normalization and ReLU activation function. Pix2Pix learns how to convert one type of image into another. Zi2Zi [10] uses GAN to convert Chinese characters between fonts in an end-to-end manner. As shown in Figure 1(a), Zi2Zi USES pairs of Chinese characters in source and target fonts as training data. Zi2Zi's network structure is based on Pix2Pix, adding a variety of font categories embedded. This enables Zi2Zi to use a training model to convert characters into several different fonts. However, since it is impractical to obtain a set of large number of paired training samples for the generation of personalized handwritten Chinese characters, Zi2Zi is not applicable to our problem.

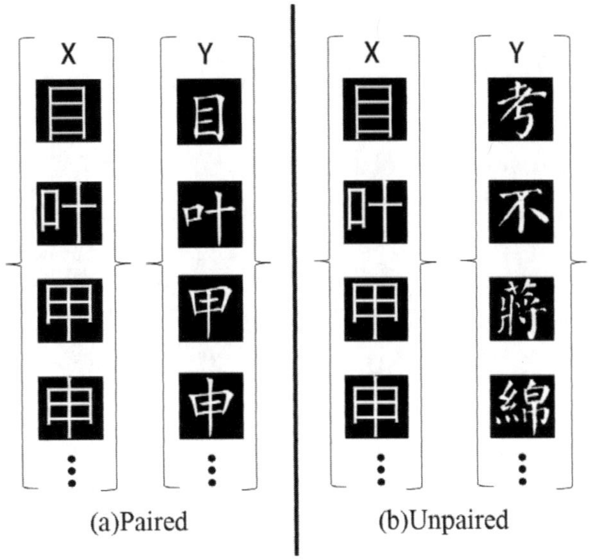

Figure 1. Paired and unpaired train data. (a) Paired training set contains training examples $\{x_i, y_i\}_{i=1}^{N}$, where there exists correspondence between x_i and y_i. (b) Unpaired training set contains training examples, where the source set X and the target set Y have no paired matching information.

3. Our method

Pix2Pix handles image conversions that match data sets very well, but in many cases, there is almost no data set that is absolutely matched, and collection is difficult. However, the amount of data that is not matched in the two fields is still very large. As shown in Figure 1(b), we can get the calligrapher's font from some inscriptions. If you can achieve image conversion through unmatched data, this work becomes very meaningful. In 2017, DiscoGAN and CycleGAN [11] proposed an image conversion scheme for solving unmatched data sets. The schematic diagram of the DiscoGAN structural framework is shown in Figure 2. Essentially two different GANs are coupled together, and each GAN maps one domain to the corresponding domain.

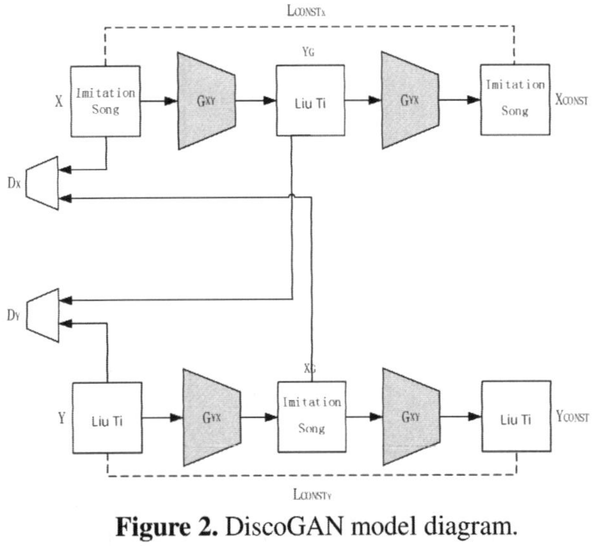

Figure 2. DiscoGAN model diagram.

In section 3.1, we introduce the network architecture adopted in this paper, and in section 3.2, we describe several loss functions adopted in this paper.

3.1 Structural framework
Figure 3 shows the network structure of the generator, which is composed of three parts: coding layer, conversion layer and decoding layer.

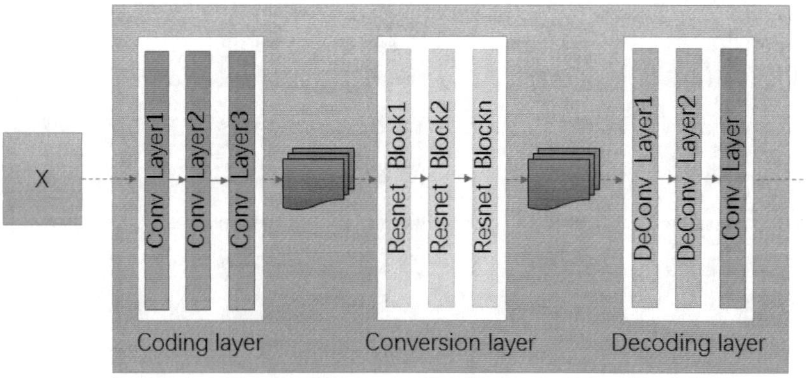

Figure 3. The generator network structure of DiscoGAN.

3.2 Loss function
The conversion between non-matching pairs of image domains is not as simple as two GANs.

$$L_G = L_{G_{XY}} + L_{G_{YX}} = L_{GAN_Y} + L_{CONST_X} + L_{GAN_X} + L_{CONST_Y} \tag{3}$$

$$L_D = L_{D_X} + L_{D_Y} \tag{4}$$

Where, L_{GAN_Y} is the loss of generation of Y domain in the X domain, and L_{CONST_X} is the reconstruction loss of X.

If the generator G_{X2Y} that converts X to Y expects the discriminator D_Y to discriminate the generated image as true, then G_{X2Y} preferably does not extract any information related to X, but independently learns and generates data from Y. At this time, although we got a realistic Y-domain image, this image has nothing to do with the X-domain. Therefore, a concept of reconstruction loss is introduced in DiscoGAN. In Figure 1, we combine the two sets of generated confrontation networks. Image X generates YG through G_{X2Y}, and Y_G is restored to X_{const} through G_{Y2X}. Obviously, the more consistent X and X_{const}, the closer our entire network is to our intended goal. Therefore, the distance between X and X_{const} is called the reconstruction loss:

$$\begin{aligned} L_{CONST} = (G_{X2Y}, G_{Y2X}) = \mathbb{E}_{x\sim p_{data}}(x)[\|G_{Y2X}(G_{X2Y}(x)-x)\|_1] \\ + \mathbb{E}_{y\sim p_{data}}(y)[\|G_{X2Y}(G_{Y2X}(y)-y)\|_1] \end{aligned} \tag{5}$$

4. Experiments

In this section, we will perform experimental verification on the dataset. In the experimental set, the original domain and the target domain have both common attributes and visible differences.

Song style is a kind of Chinese character printing font, which imitates the font carved in the book of song dynasty. The song dynasty typeface is people imitates the song dynasty typeface the structure, the brushwork meaning, in order to adapt to the printing, need to evolve, is very different with the writing style of the brush. The refers to the general name of the calligraphy works of Liu Gongquan, one of the last great calligraphers of the Tang Dynasty and one of the four masters of the book. The Liu Ti is balanced and thin, and the paintings are cool and pretty, the bones are strong, and the knots are tight. Compared with Yan Zhenqing's calligraphy work, it is slightly even and thin.

The training data consists of 5,000 imitations from the computer font library and 1,302 face characters (including duplicates) from the mysterious tower monument. All words are pre-processed to 256 x 256. After 200 rounds of training with our method, the results shown in Figure 4 were obtained, including many simplified characters that were not in the training set, which better preserved the style of the willow.

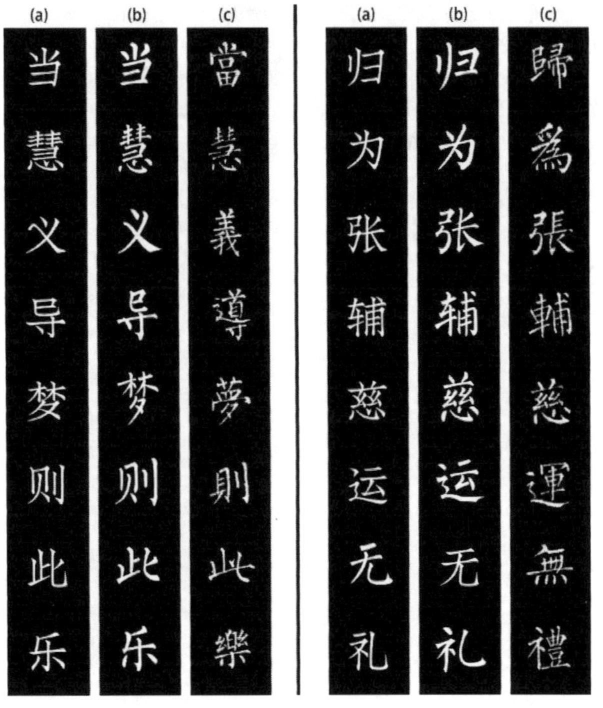

Figure 4. Partially generated word set (a) imitation Song source characters. (b) willow characters generated using DiscoGAN. (c) willow source characters.

5. Conclusion

In this work, we describe the problem of generating a style calligraphic work as a mapping problem from the existing print font to the calligrapher style. We use DiscoGAN to solve this problem. In order to make the model more suitable for our problem, the loss of identity consistency is guaranteed in the loss function, and the validity of the method is verified by experiments on the dataset. In the next work, we try to propose an evaluation function to quantify the effect of font conversion; on the other hand, we plan to study more ways of art style transfer.

References

[1] Kim T, Cha M, Kim H, et al. Learning to discover cross-domain relations with generative adversarial networks[C]. Proceedings of the 34th International Conference on Machine Learning-Volume 70, 2017: 1857-1865.

[2] Xu S, Lau F C, Cheung W K, et al. Automatic generation of artistic Chinese calligraphy[J], 2005, 20(3): 32-39.

[3] Xu S, Jin T, Jiang H, et al. Automatic generation of personal chinese handwriting by capturing the characteristics of personal handwriting[C]. Twenty-First IAAI Conference, 2009.

[4] Zong A, Zhu Y. Strokebank: Automating personalized chinese handwriting generation[C]. Twenty-Sixth IAAI Conference, 2014.

[5] Gatys L A, Ecker A S, Bethge M J a P A. A neural algorithm of artistic style[J], 2015.

[6] Goodfellow I, Pouget-Abadie J, Mirza M, et al. Generative adversarial nets[C]. Advances in neural information processing systems, 2014: 2672-2680.
[7] Isola P, Zhu J-Y, Zhou T, et al. Image-to-image translation with conditional adversarial networks[C]. Proceedings of the IEEE conference on computer vision and pattern recognition, 2017: 1125-1134.
[8] Mirza M, Osindero S J a P A. Conditional generative adversarial nets[J], 2014.
[9] Radford A, Metz L, Chintala S J a P A. Unsupervised representation learning with deep convolutional generative adversarial networks[J], 2015.
[10] Tian Y. zi2zi: Master chinese calligraphy with conditional adversarial networks, 2017.
[11] Zhu J-Y, Park T, Isola P, et al. Unpaired image-to-image translation using cycle-consistent adversarial networks[C]. Proceedings of the IEEE International Conference on Computer Vision, 2017: 2223-2232.

Approximate Yield Estimation of Correlated Failure Events for Near-threshold SRAM

Zhongmao Sun*, Shuaibo Huang and Hao Yan

National ASIC System Engineering Research Center, Southeast University, No.2 Sipailou, Nanjing, Jiangsu Province, China

*E-mail: szm@seu.edu.cn

Abstract. In advanced manufacturing technology, SRAM yield estimation with correlation is a significant challenge due to process variations and structures especially at low voltage. In this paper, Double Asymptotic Probability Approximation (DAPA) method with Subset Analytical Model (SAM) is proposed to effectively estimate the probability of correlated failure event. SAM models SRAM failure events based on simulation of subsets of the equivalent circuit. DAPA obtains the recursive function of partial failure rate. The key idea is to approximate the failure rate of the entire system based on a linear combination of partial failure rates by decoupling failure correlation. Under TSMC 28 nm process, ($M=N=8$) SRAM failure rate is estimated at 500 mV with an error of less than 5% by the proposed method. Besides 2.5 hours of modeling consumption, the calculation can be completed in 7 seconds, which greatly reduces SRAM failure estimation time, compared with Monte Carlo method.

1. Introduction

The on-chip SRAM is commonly used in SoC systems and consumes a significant fraction of the total chip power. For high integration density, SRAM has always been designed with the advanced integrated circuit (IC) technology which brings about serious process variations, making SRAM bitcells extremely vulnerable to failures especially at low voltage caused by random dopant fluctuations, litho-induced variations, oxide thickness variations, etc. However, as the supply voltage approaches the threshold voltage of transistors in CMOS technology, energy efficiency is near to the optimal point [1]. And near-threshold SRAM is widely applied in low-power embedded systems and IoT platforms. Thus, to achieve required or aggressive performance and power, it is contradictory to guarantee sufficient yield. Fortunately, some applications (e.g., neural networks) have been shown to be inherently error resilient, which may perform under the presence of bit errors to further improve energy-efficiency based on effective yield estimation.

Statistical methods addressing yield problems have been proposed in the literature. A widely used approach is Monte Carlo (MC) method based on randomly selected samples in the variation space. However, MC typically requires millions, or even billions of samples to reach reasonable accuracy, which is prohibitively costly. Therefore, other approaches based on importance sampling (IS) [2] and boundary-searching [3] are developed. The basic idea of IS is to sample the failure regions with distorted probability density function. And boundary-based methods are to construct the boundary of the failure regions. However, these approaches can only deal with low-dimensional variation spaces. Subset Simulation (SUS) [4] is proposed to handle the high-dimensional problem, which estimates the failure rate by multiplying the probabilities of several intermediate failure events.

The aforementioned methods only focus on the cell-level failure events (i.e., single bitcell failure) which cannot reflect the actual failure rate of entire SRAM. However, it is unrealistic to run Monte Carlo simulations of a large SRAM for the computational complexity and time consumption. Failure rate estimation of entire system has not been intensively studied yet. The difficulty is that system-level failure events are determined by a large number of cell-level failures. And the cell-level failures are often statistically correlated due to circuit structures [5], global process variations [6], supply voltage variations, etc. However, most existed methods assume that cell-level events are independent. In this case, ignoring or inappropriate simplifying correlation will lead to significant inaccuracy in yield estimation for relatively large SRAM arrays [7]. Analytical model [8] reflects the correlation among bitcell failures caused by the shared peripherals. But, performance parameters are modeled as Gaussian distributions which may not be consistent with actual distributions in advanced manufacturing technologies and at low voltage [9]. A novel Asymptotic Probability Approximation (APA) [10] is proposed to estimate system-level failure rate without distribution assumption that is approximated as a linear combination of partial failure rates. And the partial rates are calculated by Hierarchical Subset Simulation (H-SUS) method. But, the variation of sense amplifier enable time is not considered.

In this paper, the system-level failure events with correlation among bitcells are estimated with Double Asymptotic Probability Approximation (DAPA) and Subset Analytical Model (SAM). DAPA method is to decouple the system-level failure into a set of column-level, line-level and partial failure events. Then system-level failure rate is approximated as a linear combination of column-level, line-level and partial failure rates in two different ways. And it assumes they are approximately equal which can obtain the recursive function of partial failure rate. That is very different with APA. To make DAPA of practical utility, SAM is established by subsets of the equivalent circuit simulation and further decouples the correlation among bitcells. And Look-up Table (LUT) is utilized to record the relationship among performance parameters. Additionally, subsets of the equivalent circuit are not very dependent on system circuit, which means the simulation experiments can be done beforehand.

2. Background

2.1. SRAM Failure Model

Process parameter variations result in the mismatch in the strengths of different transistors in a single SRAM cell, which leads to functional failures (read, write, access and hold failure) [11]. Assume that the process variations are described by the vector $\boldsymbol{x}$ of a series of independent random variables and $y^{cell}(\boldsymbol{x})$ denotes the performance of interest (PoI) of a single SRAM bitcell. The cell-level failure rate can be expressed as:

$$p^{cell} = \Pr\left[y^{cell}(x) \in F^{cell} \right] \tag{1}$$

where F^{cell} represents the cell-level failure region, and $\Pr[\cdot]$ represents the probability of an event [10]. Traditionally, F^{cell} is estimated by Monte Carlo (MC) simulation by studying cell-level PoI with N random samples, $x^{\{i\}}$ $(i = 1, 2, ..., N)$, generated by its probability density function (PDF):

$$p^{cell} = \frac{1}{N} \cdot \sum_{n=1}^{N} I^{cell}\left[y^{cell}(x^n) \right] \tag{2}$$

where $y^{cell}(x^n)$ stands for the value of cell-level PoI on the sample x^n, and $I^{cell}[\cdot]$ is the indicator function. For instance, the read access failure can be modeled as the statistical relation among sense amplifier (SA) offset V_{OS}, sense amplifier enable time T_{SAE}, and the voltage difference between complementary bitlines ΔV_{BL} [8]. The dominant reason for read access failure is that a large V_{OS} or a short T_{SAE} makes ΔV_{BL} smaller than V_{OS} . Moreover, the parameters of failure model are all unknown distributions due to process variations, especially at low supply voltage. Thus, the $I^{cell}[\cdot]$ of read access failure is:

$$I^{cell}\left[y^{cell}(x)\right]=\begin{cases}1 & \Delta V_{BL}(t)<V_{OS}\\ 0 & \Delta V_{BL}(t)\geq V_{OS}\end{cases} \tag{3}$$

where t is drawn from PDF of T_{SAE} . The failure mechanism becomes complex, as the $\Delta V_{BL}(t)$ is an uncertain distribution at any time. And the dimension of parameter space will increase, as all the transistors are supposed to be taken into consideration. Thence, a meaningful failure estimation requires millions, or billions of samples to capture effective failure regions. In theory, to estimate the probability P with (1-ε) accuracy and (1-δ) confidence, according to [12], the required sampling number of MC method is:

$$NUM^{MC}=\frac{\log(1/\delta)}{P\cdot\varepsilon^{2}} \tag{4}$$

2.2. Correlated Failure Events

The system-level failure events are determined by a large number of cell-level failures. And the cell-level failures are often statistically correlated due to circuit structures [5], global process variations [6], supply voltage variations, etc. However, most existed methods assume that cell-level events are independent. In this case, ignoring or inappropriate simplifying correlation will lead to significant inaccuracy in yield estimation for relatively large SRAM arrays [7].

SRAM array is shown in Figure 1, which consists of N columns with a shared replica bitline circuit. And each column consists of M bitcells that share the same SA. Before a read operation, complementary bitlines (*BLs & BLBs*) have been precharged to supply voltage and wait for read enable signal. Then corresponding word line *WL[0]* and *RWL* are enabled and the *RBLB* begins to discharge through the replica cell. At the same time, bitlines are discharging through the selected bitcells and the ΔV_{BL} is increasing. If SAE is enabled early when the developed ΔV_{BL} is smaller than the SA offset voltage, a read access failure occurs and bitcells of the same line may fail. Additionally, if V_{OS} is relatively higher, bitcells of the same column are highly probable to fail. Notice that due to the shared SA and replica bitline circuit, cell-level failure events are correlated.

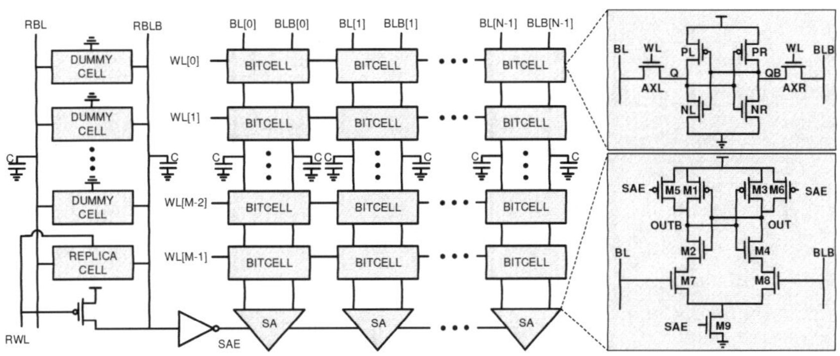

Figure 1. SRAM array with conventional replica bitline technique.

Furthermore, the distribution of T_{SAE} is non-Gaussian [9] with a long tail because of weak bitcells at near-threshold voltages. If SAE is not enabled during word line enable time, the system-level failure occurs as well. Therefore, it is contradictory to guarantee sufficient yield to achieve required performance.

3. Proposed Approach

3.1. Problem Definition

In this work, we focus on estimating near-threshold SRAM yield based on error resilient applications as power or performance is susceptible to process variations at low voltage. And we are not faced with the rare failure event (10^{-8}~10^{-6}) problem.

Suppose that we have a system containing *N* columns with a shared replica bitline circuit. And each column consists of *M* bitcells that share the same SA. The system-level failure event occurs if any one of the bitcells fails. Let $\boldsymbol{x}$ denotes its process-level random variables and involves global and local variations. The system-level PoI can be expressed as $\boldsymbol{y(x)} = [y_{11}(\boldsymbol{x})...y_{1N}(\boldsymbol{x})...y_{M1}(\boldsymbol{x})...y_{MN}(\boldsymbol{x})]$, where $y_{MN}(\boldsymbol{x})$ means the PoI of the *N*-th bitcell of *M*-th line. If PoI of any bitcell does not satisfy the specification, system fails. To describe system-level failure events and the correlation issues, the conception of set is introduced. The system-level failure event can be expressed as a union of all cell-level failure events and the failure rate P^{ARR} can be expressed as:

$$P^{ARR} = \Pr[y(x) \in F^{ARR}] = \Pr[cell_{11} \bigcup ... \bigcup cell_{1N} ... \bigcup cell_{M1} ... \bigcup cell_{MN}] \tag{5}$$

Due to the correlation of cell-level failure events, $cell_{ij} \bigcap cell_{uv} \neq \varnothing$, where $cell_{ij}$ represents cell-level failure event $y^{cell_{ij}}(x) \in F^{cell_{ij}}$, and $ij, uv \in \{11, 12, ..., MN\}$. But, the failure rate of different bitcells is regarded to be same as all bitcells perform the same function [10]. If we completely ignore the correlations of cell-level failures (i.e., assume the failure events of different cells are independent), system-level failure estimation can be calculated by binomial distribution:

$$P^{ARR} = 1 - (1 - P^{cell})^{MN} \tag{6}$$

However, this method will lead to significant inaccuracy especially when *MN* is large. To accurately estimate system yield, effective analytical models considering correlation are required.

3.2. Double Asymptotic Probability Approximation

We propose the Double Asymptotic Probability Approximation (DAPA) method to decouple the system-level failure into a set of column-level, line-level and partial failure events. Then system-level failure rate is approximated as a linear combination of column-level, line-level and partial failure rates in two different ways. The partial failure events are defined as:

$$P^{mn} = \Pr[cell_{11} \bigcup ... \bigcup cell_{1n} ... \bigcup cell_{m1} ... \bigcup cell_{mn}] \tag{7}$$

where *m<M, n<N,* and $P^1 = P^{11}$. The partial failure rate can also be expressed by column-level or line-level failure rates:

$$P^{mn} = \Pr[col_1 \bigcup col_2 ... \bigcup col_n] = \Pr[line_1 \bigcup line_2 ... \bigcup line_m] \tag{8}$$

where col_n and $line_m$ are union of all cell-level failure events. Since the function of different bitcells, columns or lines is the same, the probability of partial failure events does not concern the actual bitcells, columns or lines. Thus, it can be expressed as:

$$P^{mn} = P^{m1} + P^{m(n-1)} - \Pr[col_1 \bigcap (\bigcup_{i=2}^{n} col_i)] \tag{9}$$

With the similar derivation process, $P^{m(n-1)}$ can be expressed by $P^{m(n-2)}$ and other terms. Then, P^{mn} can be expressed as:

$$P^{mn} = nP^{m1} - \sum_{k=2}^{n} \Pr[col_1 \bigcap (\bigcup_{i=2}^{k} col_i)] \tag{10}$$

In this way, P^{m1} can be represented by P^1 and other terms [10]:

$$P^{m1} = mP^1 - \sum_{k=2}^{m} \Pr[cell_1 \bigcap (\bigcup_{i=2}^{k} cell_i)] \tag{11}$$

Ultimately, P^{mn} can be expressed as:

$$P^{mn} = n(mP^1 - \sum_{k=2}^{m} \Pr[cell_1 \bigcap (\bigcup_{i=2}^{k} cell_i)]) - \sum_{k=2}^{n} \Pr[col_1 \bigcap (\bigcup_{i=2}^{k} col_i)] \tag{12}$$

Similarly, the partial failure rate can also be expressed by line-level failure rates:

$$P^{mn} = m(nP^1 - \sum_{k=2}^{n} \Pr[cell_1 \bigcap (\bigcup_{i=2}^{k} cell_i)]) - \sum_{k=2}^{m} \Pr[line_1 \bigcap (\bigcup_{i=2}^{k} line_i)] \tag{13}$$

Notice that P^{MN} can also be expressed in the form of (12), (13). The approximate relationship is established by ignoring the terms of $cell_{m+b}$ and col_{m+d} in (12), where $0 < b \leq M - m, 0 < d \leq N - n$. P^{MN} is approximated as:

$$\begin{aligned}
P^{MN} &= N(MP^1 - \sum_{k=2}^{M} \Pr[cell_1 \bigcap (\bigcup_{i=2}^{k} cell_i)]) - \sum_{k=2}^{N} \Pr[col_1 \bigcap (\bigcup_{i=2}^{k} col_i)] \\
&\leq N(MP^1 - (\sum_{k=2}^{m} \Pr[cell_1 \bigcap (\bigcup_{i=2}^{k} cell_i)] + \sum_{k=m+1}^{M} \Pr[cell_1 \bigcap (\bigcup_{i=2}^{m} cell_i)])) \\
&-(\sum_{k=2}^{n} \Pr[col_1 \bigcap (\bigcup_{i=2}^{k} col_i)] + \sum_{k=n+1}^{N} \Pr[col_1 \bigcap (\bigcup_{i=2}^{n} col_i)]) \\
&= NMP^1 - N(mP^1 - P^{m1} + (M-m)(P^1 + P^{(m-1)1} - P^{m1})) - (nP^{m1} - P^{mn} + (N-n)(P^{m1} + P^{m(n-1)} - P^{mn})) \\
&= N(M-m)(P^{m1} - P^{(m-1)1}) + (N-n)(P^{mn} - P^{m(n-1)}) + P^{mn} = Q^{mn}
\end{aligned} \tag{14}$$

In the same way, P^{MN} can also be approximated by (13):

$$P^{MN} \leq M(N-n)(P^{1n} - P^{1(n-1)}) + (M-m)(P^{mn} - P^{(m-1)n}) + P^{mn} = R^{mn} \tag{15}$$

It can be mathematically proven that Q^{mn} and R^{mn} are upper bounds of the actual failure rate. The relationship between Q^{mn} and a subset of Q^{mn} or between R^{mn} and a subset of R^{mn} can be easily determined by applying set theory.

$$P^{MN} \leq Q^{mn} \leq Q^{(m-1)n} \leq Q^{(m-1)(n-1)}, P^{MN} \leq R^{mn} \leq R^{m(n-1)} \leq R^{(m-1)(n-1)} \tag{16}$$

Assume that Q^{mn} and R^{mn} are approximately equal since they all represent the upper bound of the actual failure rate. P^{mn} recursive function can be expressed as:

$$P^{mn} = \frac{M}{1-t}(P^{1n} - P^{1(n-1)}) - \frac{tN}{1-t}(P^{m1} - P^{(m-1)1}) + \frac{P^{m(n-1)}}{1-t} - \frac{tP^{(m-1)n}}{1-t} \tag{17}$$

where $t = (M-m)/(N-n)$. $P^{m(n-1)}$ and $P^{(m-1)n}$ can be expended in this way. At length, P^{mn} is expressed by a set of column-level and line-level failure rates. To estimate $P^{(m-j)1}$ and $P^{1(n-j)}$, the simultaneous failure rate $P_s^{(m-j)1}$ and $P_s^{1(n-j)}$ are introduced which mean the probabilities of all bitcells fail at the same time. $P^{(m-j)1}$ and $P^{1(n-j)}$ can be written as [10]:

$$P^{(m-j)1} = \sum_{k=1}^{m-j}(-1)^{k-1}\binom{m-j}{k}P_s^{(m-j)1},\ P^{1(n-j)} = \sum_{k=1}^{n-j}(-1)^{k-1}\binom{n-j}{k}P_s^{1(n-j)} \tag{18}$$

Hence, the remaining problem is how to effectively estimate $P_s^{(m-j)1}$ and $P_s^{1(n-j)}$.

3.3. Subset Analytical Model

The Subset Analytical Model (SAM) is proposed to solve the remaining problem. According to the definition of P_s^L, where L represents any size of column or line, P_s^L can be written in the form of conditional probability:

$$P_s^L = \Pr[cell_1 \bigcap cell_2 ... \bigcap cell_L] = \Pr[cell_1]\prod_{i=2}^{L}\Pr[cell_i \mid cell_1,...,cell_{i-1}] \tag{19}$$

P_s^L is the joint probability of L bitcells failure event and cell-level failure events are correlated due to the shared SA and replica bitline circuit. To address this correlation issue specifically, the read access failure event is taken as an example. From (3), we will further decouple the correlation among bitcells and P_s^L can be rewritten as:

$$P_s^L = \Pr[\Delta V_{BL}^1, V_{OS}^1, ..., \Delta V_{BL}^L, V_{OS}^L, T_{SAE}] = \Pr[\Delta V_{BL}^1, V_{OS}^1, ..., \Delta V_{BL}^L, V_{OS}^L \mid T_{SAE}]\Pr[T_{SAE}] \tag{20}$$

The cell-level failure model is established with three parameters (ΔV_{BL}, V_{OS}, T_{SAE}). According to the circuit structure shown in figure 1, each column shares the same SA and replica bitline circuit. However, each line only shares the same replica bitline circuit. Thus, column-level failure events depend on V_{OS} and T_{SAE}. Line-level failure events only depend on T_{SAE}. ΔV_{BL} is independent among different bitcells and V_{OS} is independent in line-level failure events. $P_s^{(m-j)1}$ and $P_s^{1(n-j)}$ can be rewritten in different ways as:

$$P_s^{(m-j)1} = \Pr[\Delta V_{BL}^1, ..., \Delta V_{BL}^{m-j}, V_{OS}, T_{SAE}] = \Pr[\Delta V_{BL}^1, ..., \Delta V_{BL}^{m-j} \mid V_{OS}, T_{SAE}]\Pr[V_{OS}]\Pr[T_{SAE}] \tag{21}$$

$$P_s^{1(n-j)} = \Pr[\Delta V_{BL}^1, V_{OS}^1 ..., \Delta V_{BL}^{n-j}, V_{OS}^{n-j}, T_{SAE}] = \Pr[\Delta V_{BL}^1, V_{OS}^1 ..., \Delta V_{BL}^{n-j}, V_{OS}^{n-j} \mid T_{SAE}]\Pr[T_{SAE}] \tag{22}$$

The conditional probability $\Pr[cell_i \mid cell_1,...,cell_{i-1}]$ in (19) is on the condition of T_{SAE} and V_{OS} or just T_{SAE} in different cases analytically. To calculate the conditional probability, plenty of samples of ΔV_{BL}, V_{OS} and T_{SAE} are generated by simulation and Look-up Table (LUT) is utilized for storing and managing the data. This process is not complicated since we only simulate the subsets of the equivalent circuit. The samples can be applied to all of the bitcells as they perform on the same function and we assume global variations are the same value. Let us start from the calculation of $\Pr[cell_2 \mid cell_1]$. When the samples (ΔV_{BL}, V_{OS}, T_{SAE}) of $cell_1$ are input to the failure model, the samples of T_{SAE} and V_{OS} which make $cell_1$ fail, are selected. For column-level failure events, the selected samples (T_{SAE}, V_{OS}) are combined with ΔV_{BL} of $cell_2$ itself are input to the failure model. Then we will get the failure rate of $cell_2$ on the condition that $cell_1$ fails. Notice that ΔV_{BL} is supposed to be searched in the LUT at different T_{SAE}. Then $\Pr[cell_i \mid cell_1,...,cell_{i-1}]$ can be calculated in this way.

For improving the calculation effectively, matrix is used. For instance, the discrete joint probability distribution of line-level of failure events at a determinate T_{SAE} can be expressed as:

$$\begin{pmatrix} bl_1 \\ \vdots \\ bl_m \end{pmatrix}_{m\times 1} \begin{pmatrix} os_1 & \cdots & os_n \end{pmatrix}_{1\times n} = \begin{pmatrix} (bl_1,os_1) & \dots & (bl_1,os_n) \\ \vdots & \ddots & \vdots \\ (bl_m,os_1) & \cdots & (bl_m,os_n) \end{pmatrix}_{m\times n} \tag{23}$$

where bl_m is one of the samples ΔV_{BL}, and os_n is one of the samples V_{OS}. As we can see, if we want to guarantee the sufficient accuracy, large numbers of samples are required which may consume a lot of storage resources. To solve this problem, samples are quantized for different accuracy requirements and resampled at failure regions.

Process parameter space is mapped into performance parameter space by SAM without any distribution assumption. Moreover, SAM is established by subsets of the equivalent circuit simulation, which consumes less time. Additionally, subsets of the equivalent circuit are not very dependent on system circuit. Namely, the simulation experiments can be done beforehand.

4. Implementation Details

4.1. Subset Analytical Model

The crucial conception of Subset Analytical Model is to find the dependent variables in the performance parameter space and decouple the correlation among bitcells. It makes DAPA of practical utility by giving the results of $P_s^{(m-j)1}$ and $P_s^{1(n-j)}$ in this procedure. Algorithm 1 summarizes this method.

Algorithm 1: Subset Analytical Model (SAM)

1 First, subsets of the equivalent circuit for required design are built.

2 Run simulations and obtain the discrete distributions of ΔV_{BL} $\{bl_{ij};i=1,2,...n,\ j=t_1,t_2,...,t_m\}$, V_{OS} $\{os_i;i=1,2,...,n\}$ and T_{SAE} $\{t_i;i=1,2,...,n\}$.

3 Quantize the discrete sequences of ΔV_{BL} and V_{OS} (Quantization error is from 10^{-3} to 10^{-6}).

4 Quantize the discrete sequences of T_{SAE} $\{t_1,t_2,...,t_m\}$.

5 For $t=t_1,t_2,...,t_m$

Search the LUT and obtain the samples of ΔV_{BL} at the time of t.

Keep the different samples and calculate the frequency of the same samples.

Find the overlapping region of ΔV_{BL} and V_{OS}.

Resample only at the overlapping region.

Calculate $\Pr[cell_1]=\Pr[cell_1]+\Pr[T_{SAE}(t)]\Pr[(bl_1(t),os_1),...,(bl_n(t),os_n)]$ based on (23).

Select the samples of V_{OS} and T_{SAE} that make $cell_1$ fail.

End for

6 Calculate $P_s^{(m-j)1}$ and $P_s^{1(n-j)}$ according to (19), (21), (22).

Subsets of the equivalent circuit contain bitcell, SA and replica bitline circuit. Each of them can be simulated individually with corresponding capacitances according to the extraction of post-simulation or by experience. It is necessary to generate sufficient samples to abstract the failure model from process variations and this method do not need to base on distribution assumption. The relationship

IOP Conf. Series: Materials Science and Engineering **565** (2019) 012008 doi:10.1088/1757-899X/565/1/012008

between ΔV_{BL} and T_{SAE} is based on LUT.

Because of limited memory, quantization and resampling are quite effective in step 3 and step 5. When the quantization error is less than 10^{-6}, there is almost no loss in accuracy. However, accuracy may not be accepted if the granularity of quantization is too rough.

In detail, to compare the value between ΔV_{BL} and V_{OS} , ΔV_{BL} multiplies the reciprocal of V_{OS} in the form of matrix (23) and it is easy to identify the values less than 1. These samples are in the failure regions. Then the conditional probabilities can be calculated and this procedure do not need to generate extra samples by simulation.

4.2. Double Asymptotic Probability Approximation

Double Asymptotic Probability Approximation method can be utilized to estimate system-level failure events based on SAM. This method calculates the upper bounds of P^{MN} twice and assumes they are approximately equal. P^{MN} can be approximately expressed by a set of column-level or line-level failure events eventually.

As P^{mn} is a recursive function (17), the result depends on its subset. However, $P_s^{(m-j)1}$ and $P_s^{1(n-j)}$ are already given by SAM. Thus, calculation can be step by step. We will compare the accuracy of P^{MN} under different (*m, n*) according to (14), (15).

Algorithm 2: Double Asymptotic Probability Approximation (DAPA)

1 Load $P_s^{(m-j)1}$ and $P_s^{1(n-j)}$ calculated by SAM.

2 Calculate $P^{(m-j)1}$ and $P^{1(n-j)}$ according to (18).

3 For *m = 2, 3,...(M-1)*

  For *n* = *2, 3,...,(N-1)*

    If $(N-n)-(M-m)\neq 0$

      Calculate P^{mn} based on (17).

    End if

  End for

End for

4 Calculate the upper bound of P^{MN} under different (*m, n*) according to (14), (15).

5. Numerical Experiments

The circuit in our experiments is shown in figure 1 and the bitline capacitances are all estimated 22 fF. Standard 6T SRAM cell, high-impedance input latch-type sense amplifier [13] and conventional replica bitline circuit [14] are adopted with TSMC 28nm HPC technology. The simulation experiments are designed for the read operation at 500 mV as we care more about the characteristics of near-threshold SRAM due to high energy efficiency.

It is unrealistic to run Monte Carlo simulations of a large SRAM for the computational complexity and time consumption. To verify our methods, M-by-N SRAM netlist is created (*M*=*N*=*8*). We perform a large number of numerical Monte Carlo simulations in HSPICE to estimate SRAM yield.

5.1. Subset Analytical Model

Single bitcell, SA and replica bitline circuit are simulated individually with corresponding capacitances $(22+\Delta)$ fF, where Δ is a correction term based on the extraction of post-simulation.

The number of simulation is 10^5 and the cell-level failure rate Pr[$cell_1$] is 6.697×10^{-4}. The golden cell-level failure rate is 6.678×10^{-4} according to MC simulation. $P^{(m-j)1}$ and $P^{1(n-j)}$ are estimated by MC and PIE shown in figure 2. (a). PIE means that failure events are independent calculated by (6). $P_s^{(m-j)1}$ and $P_s^{1(n-j)}$ are shown in figure 2. (b) estimated by SAM and MC. The number of bitcells varies from 1 to 8. As can be seen, SAM matches the simulation results well.

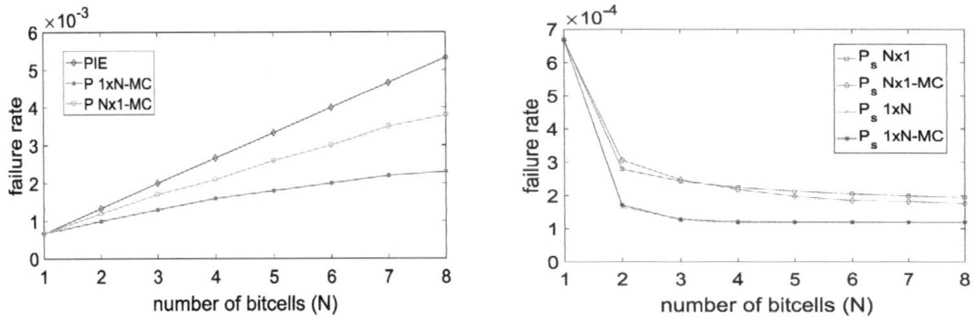

(a) **(b)**

Figure 2. (a) Failure rate estimations of column and line obtained by applying MC and PIE. (b) The simultaneous failure rate of column and line estimated by SAM and MC

5.2. Double Asymptotic Probability Approximation

P^{MN} is approximated according to (14) or (15). We compared the accuracy under different (m, n) shown in figure 3. The results of (2, 5) and (5, 2) approach the desired value well. But, when (m, n) is not appropriate, the calculation will be not very accurate. Thus, the requirements of (m, n) are analysed. If m is close to M, n is close to N or $(M\text{-}m)$ is close to $(N\text{-}n)$, corresponding terms will be eliminated according to (14), (15) and (17), which may result in inaccurate results.

Table 1 gives the numerical comparisons among MC, PIE assumption and our method. Note that the number of simulation is not quite sufficient for (M=N=8) SRAM. However, the time consumption cannot be acceptable already. Our proposed methods can be utilized to accurately estimate (M=N=8) SRAM yield with 2.5 hours of modeling consumption. And we consider the model can be built beforehand. The calculation of SAM and DAPA almost consumes no time. The error is 3.4% compared with Monte Carlo method.

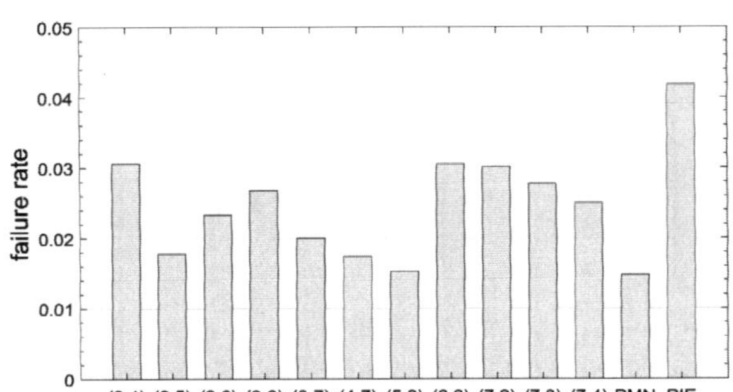

Figure 3. The failure rate of P^{MN} approximated by DAPA under different (m, n). PMN is estimated by MC. The results of (2, 5) and (5, 2) approach the desired value well.

Table 1. Numerical comparisons among MC, PIE assumption and our proposed methods.

SRAM size (M, N)	Number of simulation	Runtime of simulation	Failure rate				Consumption of our method	
			#MC	#PIE	#SAM	#DAPA		
Single bitcell	10^5	2.5 hour	6.678×10^{-4}	6.678×10^{-4}	6.697×10^{-4}	--	2.5 hour	2~5 second
(4, 4) array	10^5	34 hour	0.0054	0.0106	--	0.0049	--	2 second
(6, 6) array	10^5	76 hour	0.0099	0.0238	--	0.0095	--	2 second
(8, 8) array	10^5	134 hour	0.0148	0.0419	--	0.0153	--	2 second

6. Conclusion

In this paper, we propose Double Asymptotic Probability Approximation (DAPA) method with Subset Analytical Model (SAM) to effectively estimate the probability of correlated system failure events. SAM models SRAM failure events based on simulation of subsets of the equivalent circuit. This procedure, as we consider, can be done beforehand. Assume that the upper bounds of actual failure rate are approximately equal. DAPA obtains the recursive function of partial failure rate. The entire system failure rate is approximated based on a linear combination of partial failure rates by decoupling failure correlation. The requirements of partial failure events are analysed. Under TSMC 28 nm process, ($M=N=8$) SRAM failure rate is estimated at 500 mV with an error of less than 5% by the proposed method. Besides 2.5 hours of modeling consumption, the calculation can be completed in 7 seconds, which greatly reduces SRAM failure estimation time compared with Monte Carlo method. In the future research, we will improve the model based on importance sampling.

Acknowledgments

This work was supported in part by the National Natural Science Foundation of China under Grant NSFC 61874152.

References

[1] Alioto M 2012 Ultra-low power VLSI circuit design demystified and explained: A tutorial *IEEE Trans. Circuits Syst. I Reg. Papers* vol **59** no.1 pp 3–29

[2] Shi X, Liu F Y, Yang J and He L 2018 A Fast and Robust Failure Analysis of Memory Circuits Using Adaptive Importance Sampling Method *55th ACM/ESDA/IEEE Design Automation Conference (DAC) IEEE* pp 1-6

[3] Gong F, Yu H, Shi Y, Kim D, Ren J, and He L 2010 QuickYield: An efficient global-search based parametric yield estimation with performance constraints *Design Automation Conference (DAC)* pp 392–7

[4] Sun S and Li X 2014 Fast statistical analysis of rare circuit failure events via subset simulation in high-dimensional variation space *Proceedings of the 2014 IEEE/ACM International Conference on Computer-Aided Design. IEEE* pp 324-31

[5] Francis S, Kanj R, Joshi R, Kayssi A and Chehab A 2014 Statistical methodology for modeling non-IID memory fails events *Fifteenth International Symposium on Quality Electronic Design. IEEE* pp 205-211

[6] Wang J, Yaldiz S, Li X and Pileggi L T 2009 SRAM parametric failure analysis *Proceedings of the 46th Annual Design Automation Conference ACM* pp 496-501

[7] Zhang J and Gupta S 2014 SRAM array yield estimation under spatially-correlated process variation *IEEE 23rd Asian Test Symposium IEEE* pp 149-55

[8] Kang H, Kim J, Jeong H, Yang Y H and Jung S O 2015 Architecture-Aware Analytical Yield Model for Read Access in Static Random Access Memory *IEEE Transactions on Very Large Scale Integration (VLSI) Systems* vol **23** no.4 pp 752-65

[9] Balef H A, Kamal M, Afzali-Kusha A and Pedram M 2016 All-Region Statistical Model for Delay Variation Based on Log-Skew-Normal Distribution *IEEE Transactions on Computer-Aided Design of Integrated Circuits and Systems* vol **35** no.9 pp 1503-08

[10] Yu H, Tao J, Liao C, Su Y, Zhou D, Zeng X and Li X 2016 Efficient statistical analysis for correlated rare failure events via asymptotic probability approximation *Proceedings of the 35th International Conference on Computer-Aided Design ACM* p 18

[11] Mukhopadhyay S, Mahmoodi H and Roy K 2005 Modeling of failure probability and statistical design of SRAM array for yield enhancement in nanoscaled CMOS *IEEE transactions on computer-aided design of integrated circuits and systems* vol **24** no.12 pp 1859-80

[12] Dolecek L, Qazi M, Shah D and Chandrakasan A 2008 Breaking the simulation barrier: SRAM evaluation through norm minimization *Proceedings of the 2008 IEEE/ACM International Conference on Computer-Aided Design* pp 322-9

[13] Zhu J, Bai N and Wu J 2013 A Review of Sense Amplifiers for Static Random Access Memory *IETE Technical Review* vol **30** no.1 pp 72-81

[14] Lin Z, Wu X, Li Z, Guan L, Peng C, Liu C and Chen J 2017 A pipeline replica bitline technique for suppressing timing variation of SRAM sense amplifiers in a 28-nm CMOS process *IEEE Journal of Solid-State Circuits* vol **52** no.3 pp 669-77

The Generalized Modular Design Method of Emergency Rescue Equipment Vehicle

Li Anhu[1], Li Qiao[1], Gao Fei[1], Zhou Yihe[2], Liu Guangjun[1*] and Li Mengru[1]

[1]School of Mechanical Engineering, Tongji University, Shanghai, 201804, China
[2]China Ocean Engineering Shanghai Company, Shanghai, 200090, China

[*]E-mail: gjliu@126.com

Abstract. Emergency rescue equipment vehicles (EREV), with the ability that carry a variety of rescue equipment in different rescue scenes, play an important role in emergency rescue field. However, there still exists some issues in existing EREV products, such as small types and quantity, large volume, small loading capacity, low design efficiency and so on, which can not meet the increasingly complex rescue needs. Because of its low cost, high efficiency, strong interchangeability and reliable product performance, the modular design technology is widely used in machine tool manufacturing, toy production, electronic products, aerospace, automobile manufacturing, building construction and other fields. In this paper, a process model of the generalized modular design method for EREV is designed in order to solve the above issues. All structures and functions of modules in a product genealogy are expressed in a matrix form. This generalized modular design method proposed in this paper gives theoretical references to the design issues of other similar construction machinery devices.

1. Introduction

The emergency rescue equipment vehicle (EREV), as a carrier of all kinds of rescue equipment, plays a vital role in enlarging the function of rescue equipment and improving the efficiency of rescue work. Many famous manufacturers in the market have already designed and produced EREV early. Hackney began producing EREV in 1984. Their products mainly aim at firefighting missions, special tasks, and communications commands. Pierce is a truck manufacturer who owns a wide range of rescue vehicles, and the rescue vehicle types mainly include walk-in type, non-walk-in type and combined type. Walk-in means that rescuers can enter the internal of the vehicle freely to increase the efficiency of accessing equipment. Besides, Rescue 1, the American famous manufacturer of rescue vehicle, mainly includes two series, walk-around and walk-in. Its product range is very rich and can meet the needs of customers to the utmost. But generally, there exists many issues in these rescue equipment vehicle products on the current market, such as few types, few quantities, small carrying capacity, low production efficiency and so on, which leads to the inability to meet the complex and varied rescue needs.

In terms of the modular design technology, it is currently in a state that practice is ahead of theory. There has been a long-term infiltration of modular ideas in the fields of machine tools, reducers, computers, household appliances and furniture, but most fields are at the initial stage and a relatively systematic theoretical system has not yet been formed. Therefore, the modular design technology cannot solve the basic contradiction of product diversification and low cost for now. Many research scholars have already begun the theoretical supplements and practical research works in modular design. In the method of modular design, Gao Suying [1] used a similar feature clustering method to form generalized modules combined with the corresponding design theory, and the roving frame was expressed in the generalized module form and the generalized modular prototype system was constructed. Tang Tao et al. [2] incorporated the green design idea into the modular design, and introduced the green modular design method in detail. In the field of modular design, the module partition method is a key technology in modular design and has always been a research hotspot. Chen Yanhui et al [3] summarized the existing module partitioning methods, and classified these methods into three types roughly: function based, function and structure based, and life-cycle-based. In order to solve the high cost problem of customized product diversification, Hao Li et al. [4] proposed an interactive modular design process by analyzing the relationship between the physical module and the service module of integrated service products. The module partition principle of integrated service products is dealt with and the three-stages module partition processes and methods are put forward. A case study using an electric transformer was carried out to prove that the methods can better meet the customer's physical product and service requirements. Based on the analysis of geometric constraints, mechanical strength, energy flow, information flow and other related types, Yuo-Tern Tsai et al. [5] uses the related concepts of concurrent engineering to perform fuzzy clustering on the functional correlation. From the perspective of green life cycle, Hwai-En Tseng et al. [6] proposed a method to evaluate the product modularization. Firstly, the correlation density between products is evaluated by using an integral system, then the product modules are clustered by the cluster genetic algorithm. And finally, the clustering results are evaluated by pollution level and cost.

At present, the modular design method is relatively mature, and a large number of practical and reliable modular products have been produced by this method. However, most scholars have not systematically studied the modular methods, and some module partition methods and module expressing methods are cumbersome and not universal. Therefore, a modular process model of products is designed systematically in this paper in order to realize the basic functions of products. In this model, the modules in a product series are all expressed by a matrix. In addition, by adding special modules to the product platform, various types of EREV, which are applicable to different rescue scenarios, can be designed quickly and efficiently.

2. Generalized modular design method

2.1. Design of the process model

The generalized product modular design method is proposed in this paper, the detailed process is shown in Figure 1. Firstly, market demands or engineering issues are investigated and analyzed to obtain the overall function of products, which is then partitioned to be functional units step by step. Secondly, generalized modules are formed through incorporating functional modules with the information of structures and parameters. Then, the product genealogy is constructed and analyzed. Combined with the concept of generalized module, different structures can be designed quickly by using the parametric design technique. At last, a module database, contained information of structures, functions, parameters and so on, is built and integrated into a digital design platform for quick designing of a desired product.

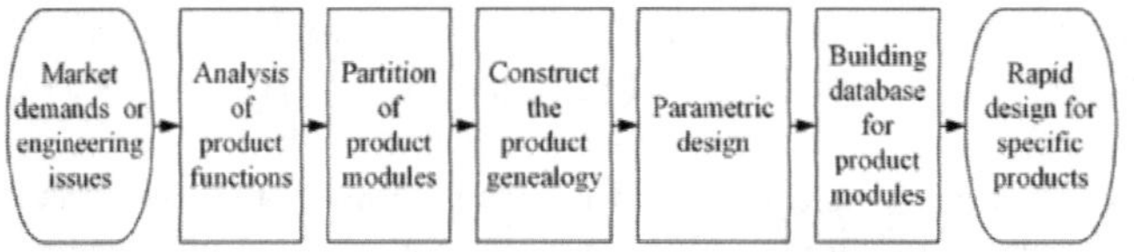

Figure 1. The process model of the generalized modular design method.

2.2. The analysis of product functions

With the increasing demand of users and the improvement of Engineering technology, the structure and function of industrial products are becoming more and more complex. In order to meet market demands and solve one type of engineering issues, the overall function of products should firstly be determined. In general, the overall function of one desired product can be determined by the black box method and the abstraction method [7]. Taking the black box method as an example, a product system is regarded as an opaque black box. Under the condition that ignoring the specific structure and working principle, the operating law of a specific system is easy to be understood and the overall function of this system can be obtained by observing the changes between system inputs and outputs (there are three kinds of inputs and outputs: material, energy and signal). For simplifying complex design issues better and explaining the function of one product in detail, the overall function is necessary to be partitioned into some sub-functions. Namely, the integrated, abstract, and complex entirety is partitioned into partial, specific, and simple elements. These sub-functions are finally partitioned into functional units step by step, which are basic units that carry one special function and can't be decomposed again. These functional units are grouped into three types generally in mechanical structural design: mathematical, logical and physical [7].

2.3. Design of the generalized module and its interface

A generalized module refers to a parametric model that carries a certain function and possesses one or more structural forms. The generalized module integrates functional modules, structural modules, structure driving parameters and performance parameters, which is expressed as equation (1). Compared with the traditional module, the most notable feature of generalized modules is that the concept of driving parameters is introduced.

$$G_m = f\left(F, S, X, P_e\right) \tag{1}$$

Where, *F* is the set of functional modules; *S* is the set of structural modules, which is corresponding to the function set. One function can be implemented by one or more structural modules as mentioned above; *X* is the set of structure driving parameters. It is a key variable in product structure design and it is constrained by geometric dimension, manufacturing accuracy and so on. The structure driving parameter is independent from each other. A structural instance can be formed by driving one group of structure driving parameters that meet the requirements; In addition, *Pe* is the set of performance parameters, which is a set used to describes the module performance of material properties, load, vibration, and so on.

The connection in generalized modules is realized by module interfaces, which contain all connection information. The function of module interfaces can be defined as equation (2).

$$I = < C, T, P_a > \tag{2}$$

Where, *C* is the interface feature, which describes the geometric structures, connection forms, constraints and so on; *T* is the interface type, a Boolean variable. I belongs to the datum interface when $T = 0$ while the connection interface when $T = 1$; *Pa* is the parameter that describes the interface characteristics.

According to the definition of the module interface, the method judging whether two interfaces, $I_1 = < C_1, T_1, Pa_1 >$ and $I_2 = < C_2, T_2, Pa_2 >$, can be linked is elaborated as follows.

- (1) If $C_1=C_2$, $T_1 \vee T_2=1$ and $Pa_1=Pa_2$, I_1 and I_2 can be connected;
- (2) If $C_1=C_2$, $T_1 \vee T_2=1$ and $Pa_1 \approx Pa_2$, I_1 and I_2 can be flexibly connected;
- (3) If $C_1=C_2$, $T_1=T_2=1$ and $Pa_1=Pa2$, I_1 and I_2 can be replace by each other, but cannot be connected;
- (4) If these above conditions are not satisfied, I_1 and I_2 cannot be replaced by each other and cannot be connected.

2.4. Construction of product genealogy and platform

It is the main purpose of modular design that as many types and quantities of products as possible are assembled with as few types and numbers of modules as possible. Combined with the concept of generalized module mentioned in the above sections, it is necessary to construct a product genealogy for these products that are assembled from modules and possess many types. The product genealogy is a set of products with similar functions, and it contains many modules with complex relations [8]. All these products are assembled with the general modules and special modules. The general modules, also known as the basic modules, are indispensable and can be often reused, the structural forms and specifications of which remain unchanged or change a little. The special modules, also named as the variable modules, are customized to specific structures and specifications according to different requirements. Because products in a genealogy are similar, there are some structures that are shared in multiple modules, which is defined as the general structural module. All general structure modules can be centralized into a set, it is worth noting that this set represents the product platform. On the basis of this product platform, the corresponding products can be developed quickly by adding special modules.

3. An design example of EREV

3.1. The composition of vehicles and functional analysis

The EREV, as a special-purpose device, play a vital role in providing rescue equipment in the emergency rescue process. Figure 2 shows the typical composition of EREV. Depending on the hierarchical structure of functional units, the main functions of equipment vehicles are analyzed in this paper, as shown in Figure 3.

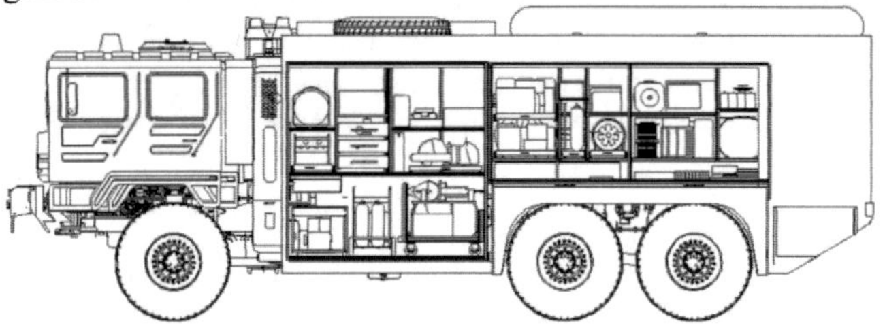

Figure 2. The typical composition of the emergency rescue vehicle.

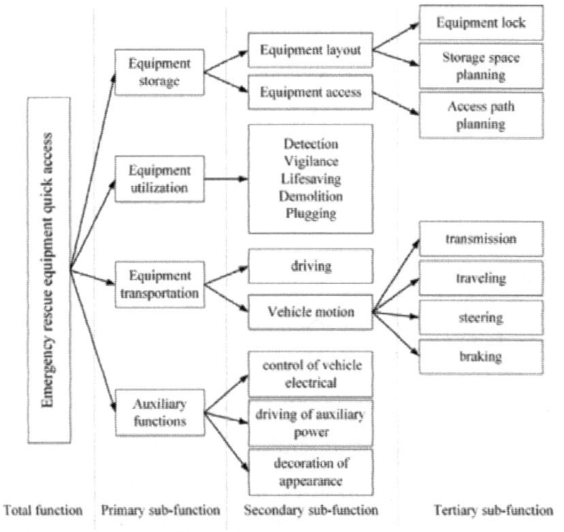

Figure 3. The functional analysis of the EREV

Generally, functions of one product are implemented, only if they are undertaken by the corresponding mechanical structures. Reasonable structural modules of one product are firstly designed according to the corresponding functional modules. Then the structural modules are assembled to each other. The overall function can be achieved finally in the form of the integral mechanical structure. According to main functions of the EREV, the mapping relationships between functional units and structural modules are researched. Many important structural modules are partitioned as shown in Section 3.2.

3.2. Generalized modules

Combined with the concept of generalized module in equation (1), major part of generalized modules is constructed on the basis of structural modules. The EREV consists of several sections of modules, and each section contains many sub-modules. Especially for modules like the special chassis, it possesses a very complex structure, and the related design methods have become mature. Detailed compositions of generalized modules in EREV are shown in Figure 4 (a). And main part of structure modules are shown as Figure 4 (b).

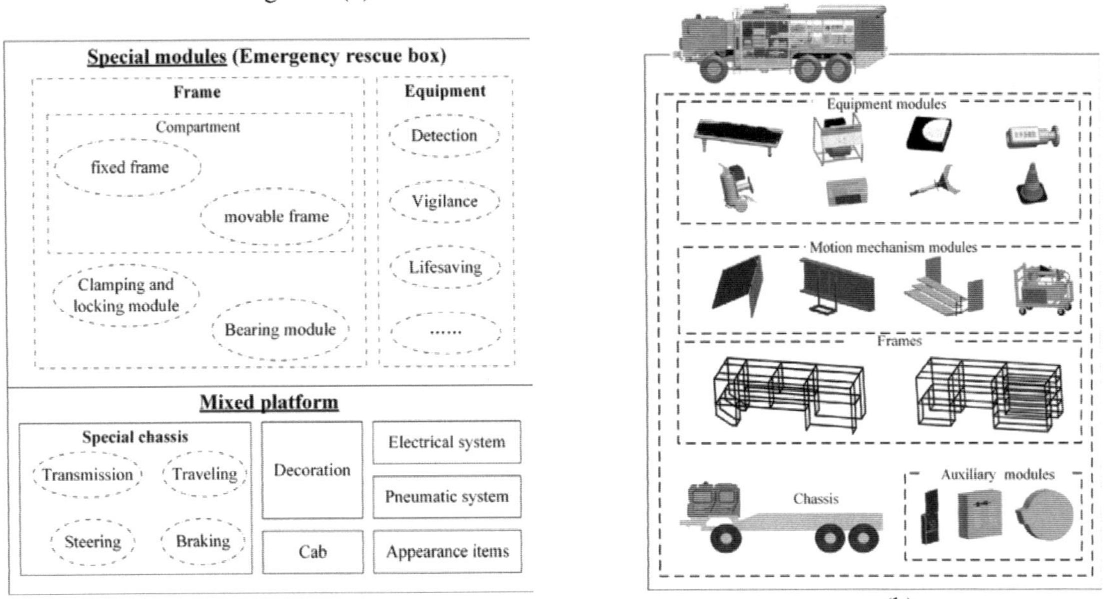

(a) (b)

Figure 4. Generalized modules of EREV: (a) main compositions, (b) structural modules

All of the modules, like frames, lock modules, bearing modules, are assembled by the module interfaces, which is the key to form an organic whole of the EREC. According to the different objects connected by the interface, module interfaces of the EREC can be divided into two categories: frame interface and installation interface. The main structures of the interfaces are shown in Figure 5. When two modules need to be connected, equation (2) can be referred to judge which type of interface should be chosen.

(a) (b)

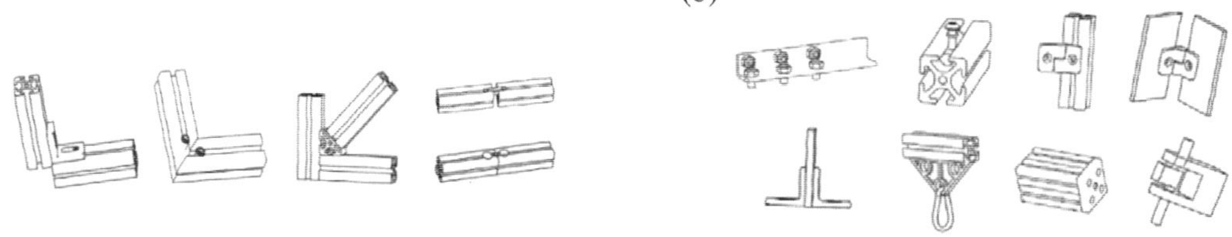

Figure 5. Structural examples of the module interfaces in the EREC: (a) frame interface, (b) installation interface.

3.3. Product genealogy and platform

Through the investigation of existing products in the market and the analysis of relevant industry standards, the product genealogy of EREV is preliminarily schemed combined with the relevant theory. According to the industry standard [9], the number of equipment on board is the main parameter of EREV. And depending on concrete quantities, EREV can be divided into three categories: small size, medium size and large size. In addition, the types of equipment that the EREC carries determine what rescue scene the vehicle aims at. Variants of the EREV are mainly applied to adopt the different rescue scenes, and referred to the industry standard [10], typical rescue scenes for EREV include: Accidents of dangerous chemicals, traffic accidents, building collapses, debris flows, floods, wind hazards, social assistances, etc. [11] Moreover, some vehicles possess special functions by equipped themselves with the on-board equipment special purposes, such as winch, on-board crane, generator, lighting system, etc. [12-13] The product genealogy of EREVs is constructed as shown in Figure 6 by analyzing the changes of specifications and functions.

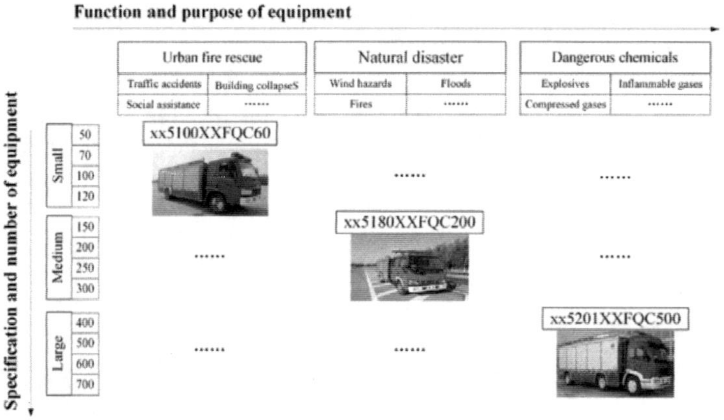

Figure 6. The serialization of the EREV product.

Among all the generalized modules of the EREV, there are strong coupling relationships between the special chassis and other modules. Almost all the structural modules will change as the structure of chassis changes greatly. In other words, the EREV is assembled and serialized on the basis of vehicle chassis. Therefore, the special chassis is the major component of the product platform. As for the modules close coupling with chassis, such as cab, electrical system, hydraulic system, appearance items and some special-purpose equipment, there are few structural variants among them, so they can also be incorporated into the product platform.

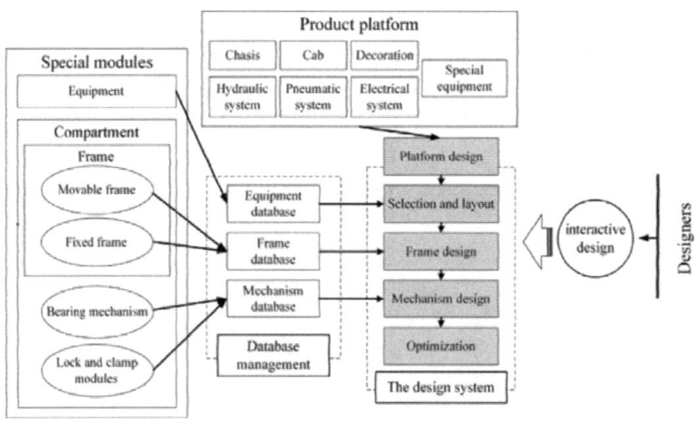

Figure 7. The software framework of the digital design platform and its working process.

The variants of the EREV are mainly determined by the types of equipment it carries, so equipment and their corresponding lock modules, bearing modules are regarded as variable modules mentioned in the Section 2.4. Besides, the compartment frame is also a variable module because its spatial structure for storage is partitioned according to the select of rescue equipment. As shown in Figure 4 (a), it not only shows the generalized modules in the EREV, but also displays the classifications of product platform and special modules.

3.4. Digital Design Platform

In this paper, the digital design platform is built according to the above analysis. As shown in Figure 7, the software framework consists of two part: the design system of EREV, the module database and its management system. The design system controls the whole design process of EREV, includes the selection and layout of equipment, the design of frames and movable mechanisms, the optimization of the EREC and so on. As for the data support for this design process, the module database plays a good job in managing the basic information relevant to the EREV modules. The design staff could expanded database by adding, deleting and editing contents through the database management system. On the basis of this database, the information in the database, about the models, properties, parameters and so on, is imported into the design system. Then, the frames, mechanisms and equipment could be easily and fast modelled and simulated. The detailed working process is also shown in Figure 7.

4. Conclusion

In this paper, the generalized modular design technology is systematically analyzed and studied, and a process model is designed. The modules in a product genealogy are all expressed in a matrix form, which is helpful to realize the horizontal or vertical design of the product. After the process model is applied to the design of EREV, the product platform, integrating most of the general modules, is established, while the module database, constructed from the parametric design technology on the special modules, is also built. Therefore, different types of EREVs can be designed quickly to meet the different engineering demand. This paper mainly focuses on the complete and systematic design method, but some key technologies, such as the module partition method, the mapping relationship between functions and structures, are not discussed in depth. In the future, we will strive to supplement and improve some of these theories.

Acknowledgments

This research was financially supported by National Key R&D Program of China (Grant No. 2016YFC0802900) and National Key R&D Program of China (Grant No. 2017YFC0307000).

References

[1] GAO S Y 2005 *Generalized modular design and software development technique of flyer frame*

[2] Tang T, Liu Z F, Liu G F, Liu L 2003 Research on the methodology of green modular design *Journal of Mechanical Engineering* **39**(11):149-154

[3] Chen Y H, Hu Y H 2013 Review of research of product modular division *Machinery Design & Manufacture* (5):264-266

[4] Li H, Ji Y, Gu X, et al. 2012 Module partition process model and method of integrated service product *Computers in Industry* **63**(4):298-308.

[5] Tsai Y T, Wang K S 1999 The development of modular-based design in considering technology complexity *European Journal of Operational Research* **119**(3):692-703

[6] Tseng H E, Chang C C, Li J D 2008 Modular design to support green life-cycle engineering *Expert Systems with Applications* **34**(4):2524-2537

[7] Liao Q L 1996 *Mechanical design methodology*. Chongqing University Press

[8] Hong H M 2013 Possibility of Application of Product Family Concept in Crane Industry *Management Observer* (24):96-99.

[9] The People's Republic of China *Specification of models for fire vehicles* GA/T 114-2003

[10] The People's Republic of China *Fire emergency rescue guidelines for equipment allocation* GB/T 29178-2012

[11] Wang J B 2007 A Study on Planning Methods of Urban Comprehensive Disaster Prevention Plan in China *Planners* **23**(1):53-55

[12] Yang B, Mei T 2014 Development Trend of Emergency Rescue Equipment in China *Labour Protection* (12):22-24

[13] Liu Q H, He T, Liu Y N. 2010 Analysis and Prospect of Emergency Rescue Vehicle Technology. *China Emergency Rescue* (5):16-17

Several Implementation Methods of Signal Processing Algorithm Based on FPGA

Wang Xin, Hu Zhi-Qiang, Jiang Da-Qing, Sun Zhi-Xiao, Zhang Yu and Liu Yong*

College of Electronic Engineering, Heilongjiang University, Harbin 150080, China

*E-mail: liuyong@hlju.edu.cn

Abstract: Field Programmable Gate Array (FPGA) devices are widely used in the field of Digital Signal Processing (DSP), and the algorithm of DSP has been used in many fields such as speech signal processing, audio signal processing, image processing, information system, control system and so on. In the development and design of modern digital systems, the use of programmable logic devices is becoming more and more common. There are many ways to implement signal processing algorithm based on FPGA: the first one is to design with VHDL or Verilog HDL language; the second one is to use mature IP core encapsulated by FPGA company; the third one is to design with EDA tools such as MATLAB fdatool/Simulink/DSP Builder/Modelsim/Quartus II. Taking the digital filter as an example, this paper uses these three methods to realize the filtering algorithm, and introduces the detailed design steps and the final simulation verification.

Keywords: FPGA; VerilogHDL; IP core; Fdatool; DSP Builder; Digital Signal Processing

1. Introduction

The rapid development of digital signal processing and related chips is closely related to digital filtering. Digital filtering has long been a research hotspot [1]. Commonly used digital filters can be divided into finite impulse response (FIR) filters and infinite impulse response (IIR) filters. Two FIR filters (Finite Impulse Response) are filters of finite-length unit impulse response. It is a non-recursive filter [2] if it is classified by recursive type.

Usually digital signal processing implementations can be roughly divided into two categories. The first type is implemented on a PC by software, and is implemented on a general-purpose computer using a computer language such as C/C++, MATLAB, and the like. This method is mostly used for teaching or algorithm simulation, and can not achieve real-time [3]. The other is board-level implementation on hardware: one is using traditional DSP processing chip programming; the other is using Application Specific Integrated Circuits (ASIC); the third is using programmable logic devices. (CPLD/FPGA). FPGA has a full parallel processing architecture, which has advantages in real-time signal processing, portable code, etc. [4], has been accepted by the majority of users, and more examples are used to implement algorithms.

FIR filters play a very important role in the application of digital signal processing because they can obtain strict linear phase, do not produce large amplitude oscillation offset in the calculation of limited precision, and their operation speed is in the digital filter. At the forefront, the advantages are favored by the majority of users.

2. FIR filter FPGA design and implementation

2.1 Theoretical basis of FIR filter
Through the filtering operation, a set of input data sequences is transformed into another set of output data sequences, thereby realizing the change of signal properties in the time domain or frequency domain [5]. A digital filter is a linear time-invariant (LTI) discrete-time system implemented by a finite-precision algorithm for implementing digital signal filtering.

The expression of the FIR filter shown in equation (1) can be expressed by a K-order convolution sum:

$$y(n) = \sum_{k=0}^{k-1} h(k) \bullet x(n-k) \quad (1)$$

Where: k: the number of taps of the FIR filter, ie the order of the filter;
$h(k)$: the k-th tap coefficient, ie the unit impulse response;
$x(n-k)$: the input signal delayed by K taps.

Convolution is one of the most frequently used basic operations for arithmetic signal processing. It describes the interaction between the system input and the system to produce an output. Typically, the convolved system output will be the input delay, attenuation, or amplification.

If the unit pulse coefficient of the filter satisfies the condition of (2) (symmetric or antisymmetric), the filter has a linear phase characteristic, ie:

$$\mathrm{h(n)} = \mathrm{h(-n)} \text{ or } \mathrm{h(n)} = -\mathrm{h(n)} \quad (2)$$

FIR filters have a variety of basic structures that are the basis for FPGA implementation. Although the FPGA is used to implement a certain FIR filter, it is necessary to adopt the corresponding implementation form of the FPGA [6]. However, no matter which FPGA implementation form is adopted, the basic structure of the FIR filter to be implemented is first determined [7], according to the basic structure of the FIR, and then proceed to choose the implementation structure of the FPGA. This article is based on different EDA tools to achieve different filter structure. In general, the basic structure of FIR can be divided into four types: direct type, cascade type, frequency sampling type and fast convolution type.

2.2 FIR filter design method
There are three methods for FIR filter design: window function method, frequency sampling method and optimization method based on firls function and remez function. The most basic design method is to require the frequency response of the designed FIR filter to approximate the frequency response required by the performance index. This paper uses the commonly used design method and window function design method in the design of FIR digital filter.

Commonly used window functions are: triangular window, rectangular window, hanning window, hamming window, Blackman window, Kaiser window. The basic parameters are shown in Table 1.

Table 1 Common window function basic parameters

Window function	Sidelobe peak/dB	main lobe width	Transition band width	stopband minimum attenuation/dB
triangularwindow	-13	4π/N	0.9	-21
rectangularwindow	-25	8π/N	2.1	-25
hanning window	-31	8π/N	3.1	-44
hamming window	-41	8π/N	3.3	-53
Blackman window	-57	12π/N	5.5	-74
Kaiser window	-57		5	-80

3 FIR filter FPGA implementation

3.1 VerilogHDL language design

There are two ideas for this design method: the first is to use the shift instead of the multiplier to achieve the multiply-add operation of the filter; the second is to use Verilog to instantiate the IP core of the associated multiplier and accumulator. The first implementation language is complex, but saves chip resources; the second is more common in engineering design, greatly simplifying the language and reducing the error rate. In this paper, a 15th-order (length 16) low-pass linear phase FIR filter is designed as an example: Blackman window function design, cutoff frequency is 500Hz, sampling frequency is 2000Hz; coefficient quantization bit is 12bit, input data bit width is 12bit The output data bit width is 26 bit system clock 2KHz.

Firstly, the filter is designed by the FDATool in the APPS (application) provided by MATLAB, and the parameters are set according to the design requirements, and the filter coefficients of the corresponding requirements can be obtained. FDATool (Filter Design & Analysis Tool) is a special filter design analysis tool in MATLAB signal processing toolbox. MATLAB6.0 and above also specifically added Filter Design Toolbox. FDATool can design conventional digital filters that meet almost all needs, including various design methods for FIR and IIR. Its operation is very simple and flexible. As shown in Figure 1, the number of filter taps, sampling frequency, passband frequency, and selection window function type are set according to design requirements, after the setting is completed, Design Filter is saved, and the coefficient output is saved to the workspace, and then the corresponding gain is added to round the output. Quantization coefficient is 0, -1, 6, 21, -52, -116, 261, 905, 905, 261, -116, -52, 21, 6, -1, 0.

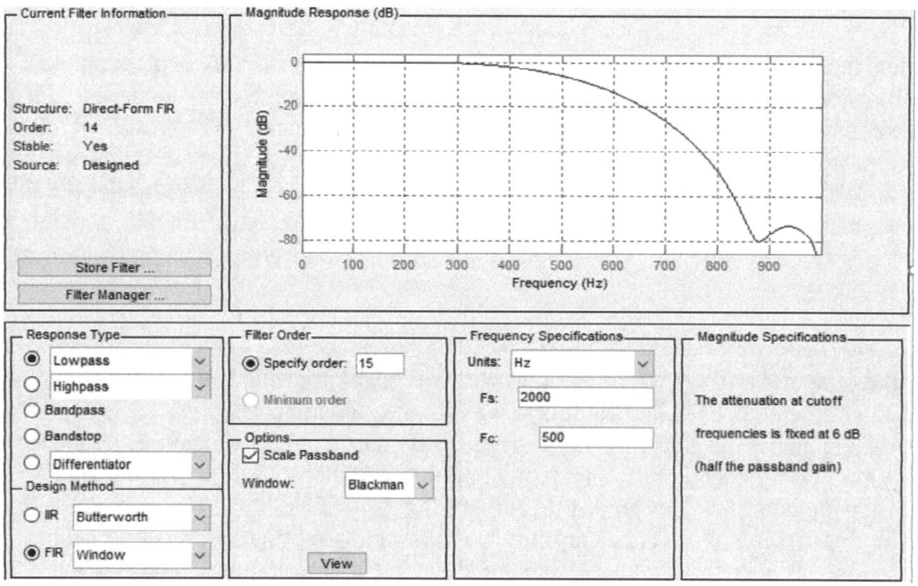

Figure 1. FDA.parameter setting

Then create a new project FIR_16 in Quartus II, select the matching hardware parameter EP4CE30F23C7; select New Design Verilog HDL file; instantiate 8 multipliers in the code (you need to compile LPM_MULT core first); manually input the above generated quantized coefficients into the multiplier in. Finally fully compiled. Check the syntax for errors and make sure the compilation is passed.

3.2 Using the FIR IP core design

According to the general PPGA design rules, for the universal functional modules implemented by the manual code, if the target device provides the corresponding IP core, the IP core is generally selected for design. The Quartus II also provides a common FIR filter IP core for most FPGA chips [8].

Therefore, in actual engineering practice, most of us will directly use the IP core provided by Altera to design the FIR filter.

Quartus Il 13.0 offers two very powerful FIR cores: FIR Compiler v13.0 and FIR Compiler II v13.0. The FIR Compiler v13.0 core data sheet details the IP core functionality and technical description. The following example of the same section above illustrates the FIR Compiler v13.0 core implementation of the distributed structure filtering algorithm.

As with Verilog's implementation of the filter, the first step still needs to generate quantized coefficients and save them as TXT format files. After opening the Quartus II software and selecting the target device EP4CE30F23C7, start the “Mega Wizard Plug-In Manager”.

Altera offers extremely rich library functions in the Mega Wizard Plug-In Manager tool, which are optimized for Altera's devices and have a simple circuit structure that reduces the designer's workload.

Select “DSP - Filters - FIRCompiler v13.0”, set the target device cluster, the language model of the output file (VerilogHDL/VHDL), determine the IP storage path and name, click “Next” to enter the IP core tool. In the interface, select “Step: Parameterize” to enter the parameter setting interface shown in Figure 2. Set the FIR core parameters. It should be noted that the bit width of the filter coefficient is set to 12 bits, the device family is CycloneIVE, and the pipeline level is (Pipline Level). The storage resources of the input data and filter coefficients are kept at default. Logic Cells. Click on the filter structure list box (Structure) to see that the IP core provides four different implementation structures: Distributed Arithmetic: Fully SerialFilter, Distributed Arithmetic: Multi-Bit Filter, Distributed Arithmetic: Fully Parall Fiter, Variable/Fixed Coefficient: Multi-Cycle. The chip resources required for different structures are different, and the operation speed is also different. For the Distributed Arithmetic: Fully Serial Filter structure option, since the filter coefficient is set to 12 bits wide (the wider the bit width, the higher the filter accuracy, the more resources are occupied); click “Edit Coeffcients Set” to configure the filter. The coefficient of the device, the IP core gives us a number of choices, we choose to automatically import the coefficients import the coefficient file of the TXT format saved in front of us.

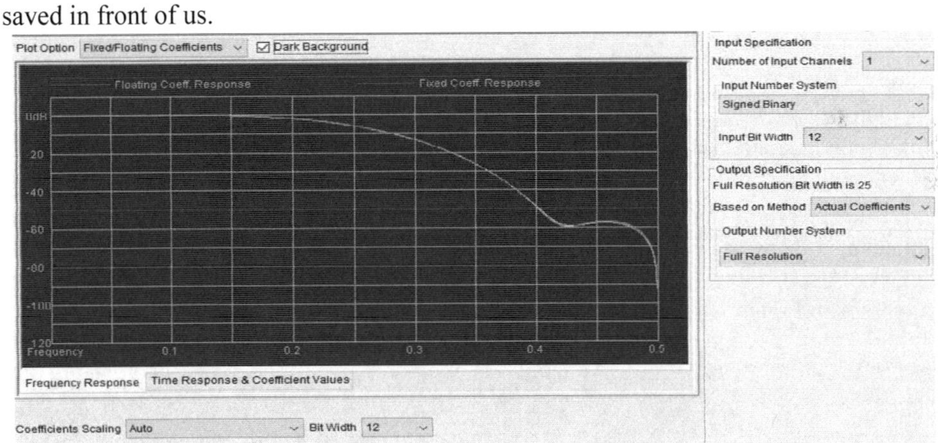

Figure 2. Filter parameter setting

3.3 Designing with DSP Builder

DSP Builder is a system-level design tool developed by Altera Corporation of the United States for DSP development. It integrates MATLAB and Simulink DSP development software in the Quartus II design environment. When co-simulation experiments on a computer, it is necessary to pay attention to the version matching problem. The matching of different versions of Quartus, DSP Builder and MATLAB can be queried on Altera's official website.

First, the filter coefficients are generated in the same way as before, and will not be described. Then create a new file, select Simulink Library, and create a new model, save it to the specified folder and name it *.mdl file.

Add Altera DSP Builder Blockset; there are two options in this step: the first option is very simple select the packaged IP core in the MegaCore Functions. As in the previous section, you can design the filter by directly configuring the IP core. The other is to build some algorithms using the integrated modules given in Arithmetic in the Altera DSP Builder Standard Blockset. The 15th-order filter uses four Multiply Add modules to divide the 16 taps (coefficients) into four groups and write them to the "Constant Values" of the four-input multiply-add module. There are also parameter settings such as the number of input bits. As shown in Figure 3.

After the core of the algorithm is packaged as Subsystem, then the peripheral circuit is built, installed as Subsystem, add counter(Increment/Decrement)and LUT (the lookup table is similar to ROM), and a Signal Compiler (linked with Quartus); double-click Signal Compiler to configure, Signaltap II Check Enable; The deeper the depth is, the more obvious the graphics will appear and the more resources will be occupied at the board level. Therefore, the depth(simulation depth) in this paper is selected as 256-2k.Finally click Export to output VHDL project.

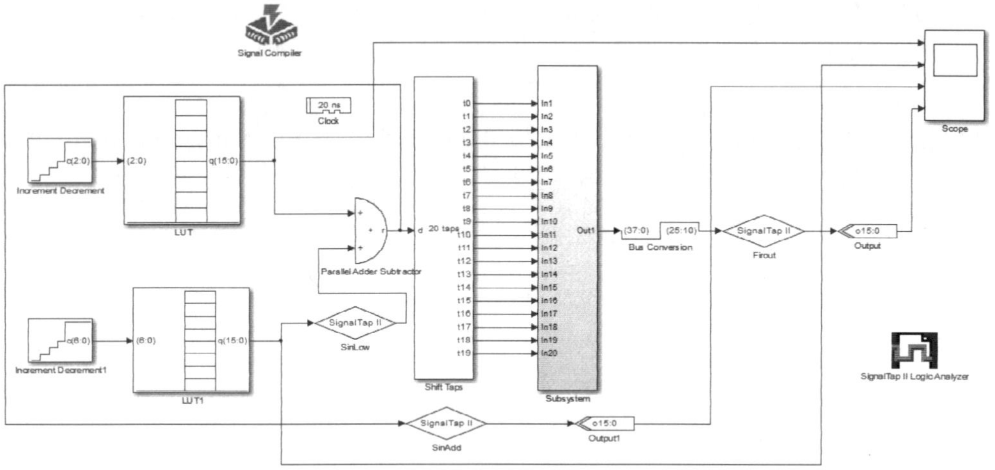

Figure 3. Overall block diagram of DSP Builder

4 Results Simulation & Conclusion

4.1 Results Simulation

Put the above three designs (take one) into the Quartus II, select Start Test Bench Template Writer in Start, generate the TestBench file, and write the testbench file. Add the testbench file to the simulation; finally click Tools, select Run Simulation Tool, select RTL Simulation, as shown in Figure 4, enter the simulation phase, and finally "dout_s" is displayed as the filtered waveform.

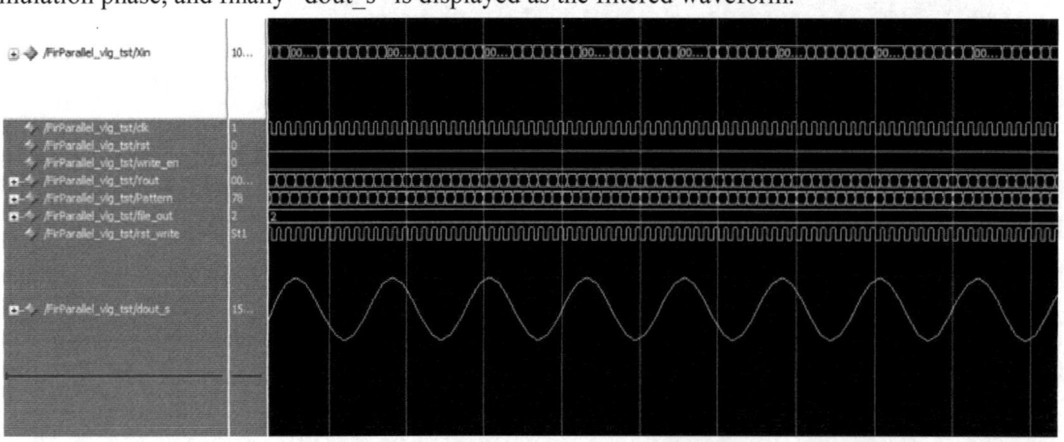

Figure 4. Simulation results of ModelSim

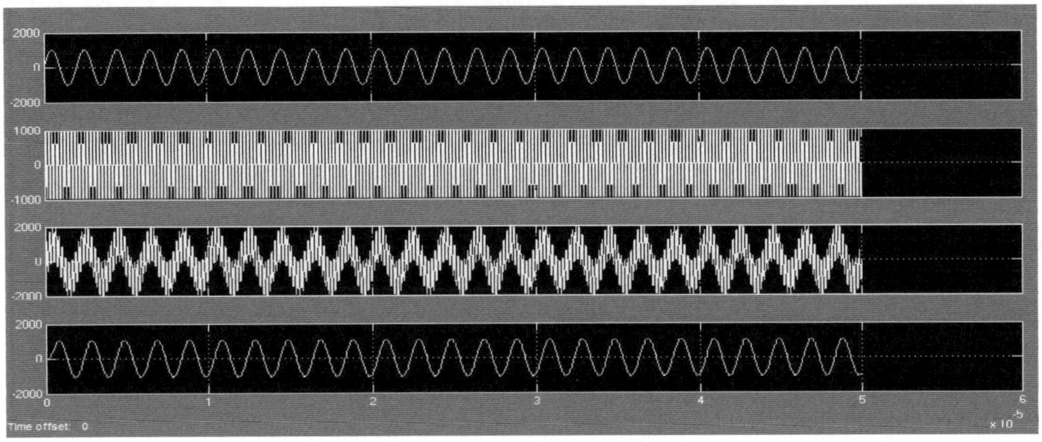

Figure 5. Simulation results of Scope

About the filtering algorithm built in DSP Builder, the simulation results can also be displayed in MATLAB's own Scope (virtual filter). As shown in Figure 5, the first window is the input sinusoidal signal; the second *window* is the superimposed high frequency signal; the third window is the accumulated mixed signal; the last window is the filtered signal.

4.2 Conclusion

Compared with DSP processing digital signals, FPGA has the advantage of saving time in parallel processing, which is not available in the latter. In the same price range, FPGA resources are much richer than DSP; in practical applications, there are many ways to implement signal processing algorithms using FPGA. Of course, in the case that the accuracy requirements are not so high, we still prefer FPGA companies to build algorithms for our encapsulated IP core and DSP Builder. The design process is simple and error-free, and further optimization of the design is needed in large projects.

References

[1] Wang Yingwei, Wang Zhenyu, Yan Wei, Shi Guangwei. FPGA Implementation and Optimization of Full Parallel FIR Filter[J]. Electronic Design Engineering, 2015, 23(22): 94-97.
[2] Yang Luxi. Modern Digital Signal Processing [M]. Beijing: Science Press, 2007.
[3] Xie Liying, Fang Limin.Design of FIR Digital Filter Based on MATLAB[J].Journal of Guangdong Second Normal University,2018,38(05):61-66.
[3] Chao Cheng, Parhi K K. Low-cost parallel FIR filterstructures with 2-stage parallelism [J]. Circuits and Systems I, 2007, 280:290.
[4] Liu Pengquan. Design and implementation of FIR digital filter based on FPGA [D]. Northwestern Polytechnical University, 2006.
[5] Chen Jiazhen, Zheng Zihua, Ye Feng, et al. Design of Digital Filter System Based on FPGA[J]. Computer Simulation, 2009, 26(12): 329-332.
[6] Dong Xianglei. Research and implementation of anti-jamming technology for navigation system based on multi-beam [D]. University of Electronic Science and Technology of China, 2014.
[7] Shan Wenjun, Zhou Xuechun, Li Wenhua.Design and Implementation of FIR Digital Filter Based on FPGA[J].Modern Electronic Technique,2013,36(14):123-126.
[8] Chen Yuanyuan, Liu Youyao. FPGA design and implementation of FIR filter[J]. Electronic Design Engineering, 2017, 25(24): 65-69+73.

Study on the Vibrant Angular Velocity Measurement Algorithm Based on Planar Sensor Array

Tianhe Zhao* **and Wei Wang**

Shijiazhuang campus of the Army Engineering University, Shijiazhuang 050003, China

*Email: 451483425@qq.com

Abstract. To solve the complex vibration testing six degree-of-freedom rigid space problem, this paper designs a sensor array configuration model and corresponding algorithm. The model integrating the method of integration and the method of the square calculates the angular velocity data. Through the analysis of the inner link of data in sensor output equation, the calculating data updates and iterations constantly, maintaining understanding of real-time and accuracy of calculation data. At the same time it can solve the problem of the error accumulation of integral method and extraction of the problem of misjudgement of symbols. The simulation results show that the error is always within the range of 10-4 in 1000 seconds.

1. Introduction

Modern platform space vibrant condition is very complex. Real-time tests and records its vibrate parameters for condition monitoring of platform, system virtual prototype designing and key components system modal analysis has very important significance [1]. The platform has many degrees of vibration freedom, large dynamic range, complex frequency components, and the vibrant state is the coupling of multi-dimensional vibration and angular vibration [2]. Unidimensional sensor testing technology can only track the movement of a single state under the condition of the vibration of the small angular velocity gyroscope responses under the condition in low frequency and it is so expensive .They are not adapted to the vibrant measurement of space [3]. The accelerometer has the characteristics of low cost, large range of testing and sensitive reaction, which provides a new way for the measurement of vibrant parameters [4]. This paper tries to configure multiple acceleration sensors in the method of plane array to detect six degrees of freedom vibrant parameters. Through the simulation analysis, we found the solution adapted to the multi degrees of freedom, large dynamic complex vibrant environment, has a good applied prospect [5].

2. Sensor array testing scheme

Installation of configurational model was shown in Figure 1. The O - XYZ is a coordinate system as establish [6]. Three triaxial acceleration sensor must be in the same position with coordinate system. The installed position of three triaxial acceleration sensors is overlapping with coordinate axis.

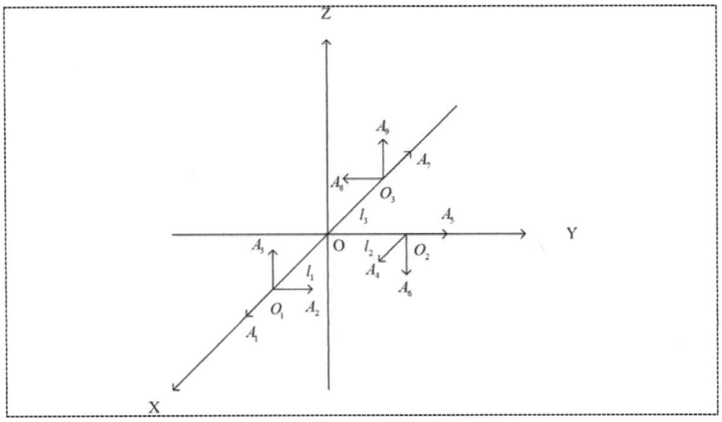

Figure 1. Schematic diagram of planar accelerometer array installation.

The positional coordinate matrix r and the sensitive direction coordinate matrix θ of 9 sensors are compared with the O - XYZ.

$$r=\begin{bmatrix} l_1 & l_1 & l_1 & 0 & 0 & 0 & -l_3 & -l_3 & -l_3 \\ 0 & 0 & 0 & l_2 & l_2 & l_2 & 0 & 0 & 0 \\ 0 & 0 & 0 & 0 & 0 & 0 & 0 & 0 & 0 \end{bmatrix}$$

$$\theta=\begin{bmatrix} 1 & 0 & 0 & 1 & 0 & 0 & -1 & 0 & 0 \\ 0 & 1 & 0 & 0 & 1 & 0 & 0 & -1 & 0 \\ 0 & 0 & 1 & 0 & 0 & -1 & 0 & 0 & 1 \end{bmatrix} \tag{1}$$

According to the kinematics equation, the output of any accelerometer based on the linear accelerometer is obtained [7].

$$f = J\begin{bmatrix} \dot{\omega} \\ \\ A \end{bmatrix} + \begin{bmatrix} \theta_1{}^T \Omega^2 r_1 \\ \vdots \\ \theta_N{}^T \Omega^2 r_N \end{bmatrix} \tag{2}$$

F --Accelerated signals measured on the accelerometers;

J -- configurational matrix;

$\dot{\omega}$ -- angular acceleration of the O - XYZ;

Ω-- the anti-symmetric matrix;

A -- the linear acceleration in the direction of the three axes of the vector coordinate system.

$$\begin{cases} \dot{\omega} = \begin{bmatrix} \dot{\omega}_x \\ \dot{\omega}_y \\ \dot{\omega}_z \end{bmatrix}^T, A = \begin{bmatrix} A_x \\ A_y \\ A_z \end{bmatrix}^T \\ \Omega^2 = \begin{bmatrix} -\omega_y^2 - \omega_z^2 & \omega_x\omega_y & \omega_x\omega_z \\ \omega_x\omega_y & -\omega_x^2 - \omega_z^2 & \omega_x\omega_y \\ \omega_x\omega_z & \omega_y\omega_z & -\omega_x^2 - \omega_y^2 \end{bmatrix} \\ J = \begin{bmatrix} J_1^T & J_2^T \end{bmatrix}, J_2 = \theta \\ r = \begin{bmatrix} r_1 & r_2 & r_3 & r_4 & r_5 & r_6 & r_7 & r_8 & r_9 \end{bmatrix} \\ \theta = \begin{bmatrix} \theta_1 & \theta_2 & \theta_3 & \theta_4 & \theta_5 & \theta_6 & \theta_7 & \theta_8 & \theta_9 \end{bmatrix} \\ J_1 = \begin{bmatrix} r_1 \times \theta_1 & r_2 \times \theta_2 & \cdots & r_9 \times \theta_9 \end{bmatrix} \end{cases} \tag{3}$$

When equation 2 was fully expanded, it can get angular velocity, angular acceleration and linear acceleration of the O – XYZ. Here is the nine accelerated sensors' equation for testing of vibration.

$$\begin{cases} f_1 = A_x - l_1(\omega_y^2 + \omega_z^2) \\ f_2 = A_y + l_1(\dot{\omega}_z + \omega_x\omega_y) \\ f_3 = A_z - l_1(\dot{\omega}_y - \omega_x\omega_z) \\ f_4 = A_x + l_2(\omega_x\omega_y - \dot{\omega}_z) \\ f_5 = A_y - l_2(\omega_x^2 + \omega_z^2) \\ f_6 = -A_z - l_2(\dot{\omega}_x + \omega_y\omega_z) \\ f_7 = -A_x - l_3(\omega_y^2 + \omega_z^2) \\ f_8 = -A_y + l_3(\dot{\omega}_z + \omega_x\omega_y) \\ f_9 = A_z - l_3(\dot{\omega}_y - \omega_x\omega_z) \end{cases} \tag{4}$$

According to the formula, the correlation expression of the related parameters represented in six degrees of freedom can be calculated.

$$\begin{cases} A_x = \dfrac{l_3 f_1 - l_1 f_3}{l_1 + l_3} \\ A_y = \dfrac{l_3 f_2 - l_1 f_8}{l_1 + l_3} \\ A_z = \dfrac{l_3 f_3 - l_1 f_9}{l_1 + l_3} \\ \omega_y^2 + \omega_z^2 = -\dfrac{f_1 + f_7}{l_1 + l_3} \\ \omega_y^2 - \omega_x^2 = \dfrac{l_3(f_2 - f_5) - l_1(f_5 + f_8) + l_2(f_1 + f_7)}{l_2(l_1 + l_3)} \\ \omega_x^2 + \omega_z^2 = \dfrac{l_3(f_2 - f_5) - l_1(f_5 + f_8)}{l_2(l_1 + l_3)} \\ \dot{\omega}_z = \dfrac{l_3(f_1 - f_4) - l_1(f_4 - f_7)}{2l_2(l_1 + l_3)} - \dfrac{f_2 + f_8}{l_1 + l_3} \\ \omega_x \omega_y = \dfrac{l_3(f_1 - f_4) - l_1(f_4 - f_7)}{2l_2(l_1 + l_3)} + \dfrac{l_3 f_9 - l_1 f_8}{l_2(l_1 + l_3)} \end{cases} \tag{5}$$

3. The design for method of angular velocity

Appling the squared of angular velocity, angular acceleration of Z axis calculating by equation 5, we can obtain three axes angular velocity by Appling to the integral method and the combination algorithm [8].

3.1. Integral method

In the case of the initial angular velocity of the O – XYZ known, the angular velocity of the O – XYZ is calculated according to the integration [9-10]. The initial parameters of the angular velocity of the O – XYZ are set, the sampling period is relatively short, and it is estimated that the time interval of the sampling interval is uniformly varying from the time interval of the sampling interval. Given the Z axis angular acceleration, the direct integral algorithm can be obtained, as shown below:

$$t = 1, 2, \cdots, n \begin{cases} \omega_z(t) = \omega_z(t-1) + \dot{\omega}_z(t)\Delta t \\ \omega_x(t) = \omega_x(t-1) + \dot{\omega}_x(t)\Delta t \\ \omega_y(t) = \omega_y(t-1) + \dot{\omega}_y(t)\Delta t \\ t = 1, 2, \cdots, n \end{cases} \tag{6}$$

The solution $\dot{\omega}_x$, $\dot{\omega}_y$ is obtained by substituting the initial parameters $\omega_x(0)$, $\omega_y(0)$, $\omega_z(0)$ .Using these parameters, the direct integral obtains $\omega_x(t)$, $\omega_y(t)$, and these parameters can use as $\omega_x(0)$, $\omega_y(0)$,etc. Angular velocity, application of the algorithm error is derived from the angular acceleration value between two sampling points considered a constant value, and cause error integral method with the extension of time to accumulate, to a large extent by increasing sampling frequency can reduce such error. Because of its inescapability, it is easy to produce large errors during the measurement of long time vibration parameters.

3.2. Open method

According to the square item in the formula, it can be obtained, but cannot directly obtain the symbol of the absolute value of three axes. In this paper, the expression of the solution of the method is expressed as the symbol sign, sign and sign, which is the absolute value of the angular velocity in the open method.

$$\begin{cases} \omega_x(t) = sign(\omega_x(t)) \times |\omega_x(t)| \\ \omega_y(t) = sign(\omega_y(t)) \times |\omega_y(t)| \\ \omega_z(t) = sign(\omega_z(t)) \times |\omega_z(t)| \\ \quad t = 1, 2, \cdots, n \end{cases} \tag{7}$$

Open method for calculation of angular velocity, process, calculating numerical not accumulate over time, avoids the production of large error, but its application is integral method calculating the value of the symbol, there is still a symbol misjudgment phenomenon, with the accumulation of time, symbol mistakes continue to increase the influence of the results.

3.3. Combination algorithm

Through analysis and comparison of various calculating methods for acceleration of plane array configurations designed an integral method, the combination of square root method algorithm, calculation of angular velocity, , the application of open method for Z axis angular velocity calculating value of the algorithm. According to related no gyro strapdown inertial navigation system alignment technology set X axis and Y axis initial angular velocity for, its plug type (4), obtain, and application, integral operation, and the integral value symbols, symbols, and its product as angular velocity calculating values. Combination algorithm based on the update of the initial value, to avoid the error integral method to accumulate over time, and continue to achieve accurate angular velocity of absolute value, through the open method combining both can obtain accurate angular velocity and settlement value of the and do not accumulate over time.

4. The simulation analysis

In order to verify the feasibility of the angular velocity algorithm for plane configuration design, the model is compared and analyzed. In the simulation, the sampling frequency is 1kHz, the sampling time is 1000s, the constant error of the acceleration sensor is 0.0001g, and the mean variance of random noise is 0.001g.

The angular velocity parameter is set to

$$\begin{cases} \omega_x = 5\sin(100t) \\ \omega_y = 5\cos(80t) \\ \omega_z = 5\sin(60t) \end{cases} \tag{8}$$

The line acceleration parameter is set to.

$$\begin{cases} A_x = 600\cos(60t) \\ A_y = 800\cos(80t) \\ A_z = 1000\cos(100t) \end{cases} \tag{9}$$

The three algorithms are simulated, and the X axis and Y-axis angular velocity errors are analyzed respectively, as shown in Figure.2 - Figure. 4.

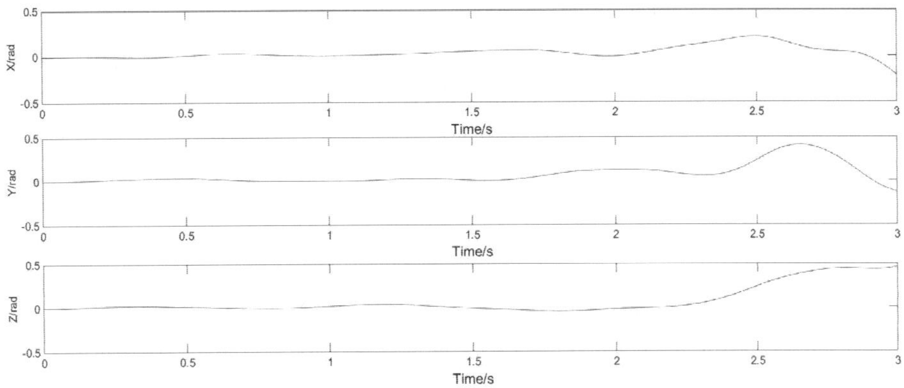

Figure 2. Integral method

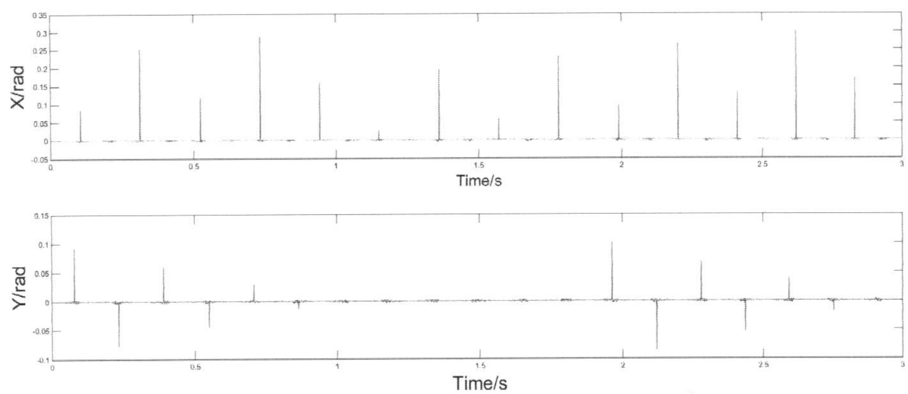

Figure 3. Open method Open method

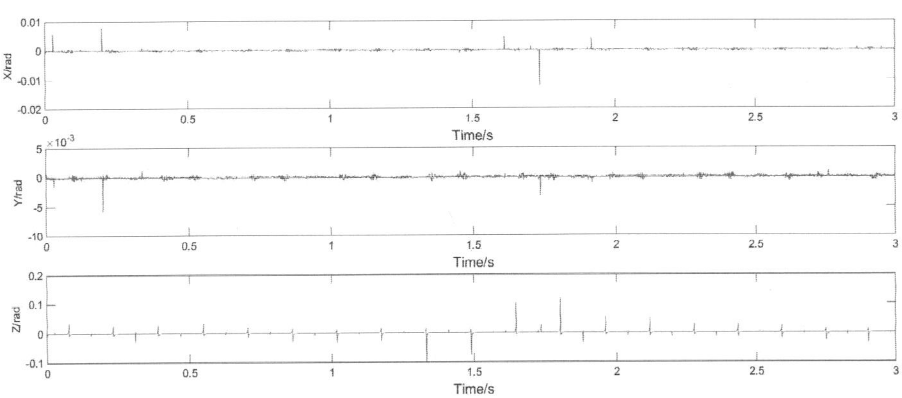

Figure 4. Combination algorithm

The angular velocity error of Figure 2 algorithm and the angular velocity error of the algorithm can be seen from the graph 3, and the error of the angular velocity is increasing over time. When the test time is short, the error is small, but when time increases, its error increases exponentially. Therefore, the algorithm cannot be applied to the attitude calculation of plane acceleration sensor array.

As can be seen from the simulation curve of Figure 3, angular velocity error will no longer increase with time accumulation to 10rad/s. Simulation shows that although open method to solve the single integral method of error accumulation problem, but the open method of the integral gain symbol component to its symbol misjudgment phenomenon with the serious time, difficult to get the real angular velocity calculating values. Therefore, it is difficult to apply integral method to new configurations.

As can be seen from the Figure 4 combination algorithm simulation, the angular velocity error for calculating combination algorithm stability in magnitude, did not appear the phenomenon of error accumulated over time, also won't appear misjudgment phenomenon, and always maintain a high accuracy of calculating results, meet the range of allowable error. The simulation shows that the combination algorithm can solve the problem of the sign misjudgment of the improved method by updating the initial value, so as to ensure the accuracy of the data. The proposed algorithm provides a reliable initial solution data for the next step, which is an effective method.

Acknowledgments

In order to realize the engineering application of sensor array technology in the field of complex vibration testing of moving body, it is necessary to design the sensor configuration scheme that meets the requirements. In this paper, a planar sensor array is introduced, and the angular velocity calculation is the key to the vibration test of sensor array. According to the new configuration, three solutions are designed by studying traditional solutions, where the error of integral method is serious with time, and the error of the method is inevitable. Combination algorithm effectively avoids the integral error accumulation and small, the angular velocity of misjudgement of symbols, in a long period of time can still maintain a high level of accuracy in the process of the simulation calculation, engineering implementation of sensor array vibration test technology has a certain significance.

References

[1] Zhao tianhe, wang wei. The current status of the MRT system based on MEMS technology [J]. Flying missile,2017(10):17-21
[2] Wang wei, jiao jianchao, tang liwei, et al. Study on the vibration test angular velocity algorithm based on sensor array [J]. China test, 2012, 38(4):58-61
[3] Jiao jianchao. Study on the test technology of complex vibration environment parameters of the fire control system. [D]. Shijiazhuang campus, army engineering university, 2011.
[4] Jiao jianchao, wang wei, tang liwei, et al. New method for vibration test of three-axis acceleration sensor array [J]. China test, 2012, 38(3):109-112
[5] Yang jie. Study and error analysis of accelerometer configuration for ungyro inertial navigation system [D]. Harbin engineering university, 2011
[6] wang wei, hao yingjie, Chen hailong, et al. Calculation of the complex vibration angular velocity of the independent kalman filter [J]. China test, 2017, 43(5):105-109
[7] Zhang hui. Self-alignment technology research and simulation [D]. Zhongbei university,2016
[8] Chen hailong. Research on the six-degree-of-freedom complex vibration measurement technology based on accelerometer array [D]. Ordnance engineering college, 2013
[9] R.Martonka,V. Fliegel. Hexapod: The Platform with 6DOF[M].Springer International Publishing:2014-06-15.
[10] Lee S C, Huang Y C. Innovative estimation method with measurement likelihood for all-accelerometer type inertial navigation system[J]. IEEE Trans on AES. 2002,38(1):339-346

IOP Conf. Series: Materials Science and Engineering **565** (2019) 012012 doi:10.1088/1757-899X/565/1/012012

The Analysis of the Negative Feedback Circuit in the Induction Sensors

Pengfei Lin[1], Chunsheng Lin[1*], Xingya Wu[2] and Ning Zhang[1]

[1]College of Weaponry Engineering, Naval University of Engineering, Wuhan 430033, China

[2]91278 unit, Lvshunkou in DaLian 116000, China

[*]Email: lcs_and_zh@163.com

Abstract. In this paper, the analysis method of the step response of the second-order system under different damping is applied to the feedback circuit analysis in the sensors. It analyzes the influence of different feedback resistors and mutual inductances on the system stability and pass band, which provides a theoretical basis for the matching of feedback resistors and mutual inductances in the circuit. It is of great significance to improve the performance of the sensor. when the parameters of the induction sensor are determined, the system performance can be more stable by designing appropriate mutual inductance and feedback resistance to make the damping ratio close to 0.8. Finally, we designed the feedback resistor Rf=400Ω, and the mutual inductance M=1 to make sure damping ratio ζ be around 0.8 in the selected sensors.

1. Introduction

Magnetic induction sensor is mainly used to measure the earth's magnetic field, its main application is the exploration of underground resources [1-2] and the weak very low frequency signal detection [3-4]. The feedback circuit of the sensor is very important to improve the system performance.

In technology research, Y. Zhao etc. present a system-theoretic approach to quantitatively model a single Aux/IAA negative feedback loop, which can be implemented deterministically and stochastically [5]. Guiyuan Wang etc. propose the analytical calculation of the stability of a core fundamental gene regulation circuit involving positive and negative interlinked regulation and time-delays [6]. David C. Marciano etc. present an alternative mechanism that is simple and uses a frequently encountered network motif. Computational and experimental evidence shows that the self-correcting, negative-feedback gene regulation motif increases repressor expression in response to deleterious mutations and thereby precisely restores repression of a target gene [7].

In this paper, it analyzes the influence of mutual inductance and feedback resistance in the negative feedback circuit on the system performance in order to design the sensors with better performance.

IOP Conf. Series: Materials Science and Engineering **565** (2019) 012012 doi:10.1088/1757-899X/565/1/012012

2. The feedback circuit

In subsequent amplifier circuits, negative feedback circuit is usually used to suppress the phase mutation and amplitude change.

2.1 Analysis of Negative Feedback Circuit

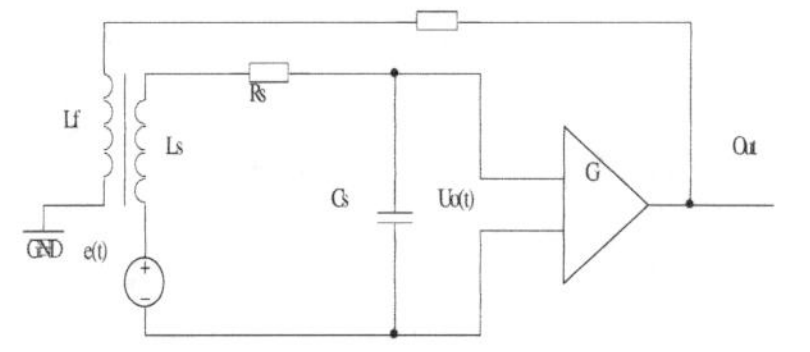

Figure 1. The negative feedback circuit diagram of the sensor.

When a negative feedback circuit is used for the sensor, the circuit is shown in figure1. The frequency characteristic function of the system is expressed as follows [8-9].

$$H(j\omega) = \frac{G}{1-\omega^2 L_S C_S + j\omega\left(R_S C_S + GM / Rf\right)}, \tag{1}$$

G is the gain multiple of the amplifier, *M* is the mutual inductance between the feedback coil and the sensor coil, and Rf is the feedback resistance. *Ls, Cs, Rs* is the distributed inductance, capacitance and the resistance of the sensors respectively. Assuming that Ls=100, $Cs=10^{-7}$F, Rs=24880Ω.

When $G = 1$, mutual inductance is fixed and Rf takes different resistance values, the Bode diagram of the characteristic function of the system is shown as follows:

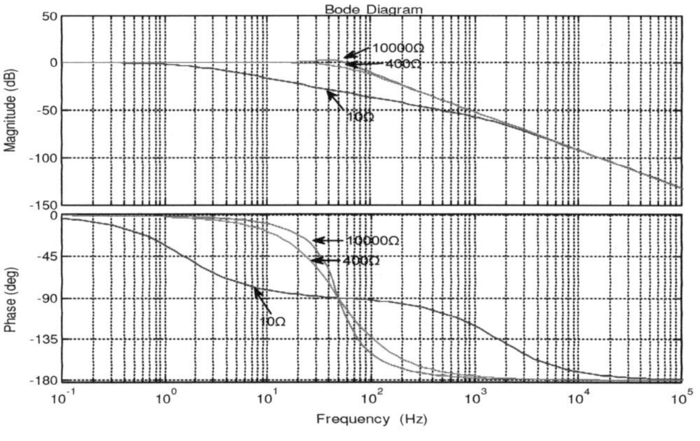

Figure 2. Bode diagram of the sensor with different feedback resistors.

The Figure 2 shows that the value of feedback resistance is very important to the performance of the sensor system. It shows that when the feedback resistance is very small to 10Ω, the pass band of the sensor decreases, and the phase-frequency changes slowly. When the feedback resistance is 400Ω, the characteristic curve of the system is similar to the system without feedback resistance, but the stability is improved. When the resistance is very large, the influence of the feedback circuit on the system becomes smaller, and it is the same as the system without feedback circuit.

When the feedback resistance is fixed, *Rf* = 400Ω, and the mutual inductance is different, the Bode diagram of the characteristic function of the system is shown in the Figure 3.

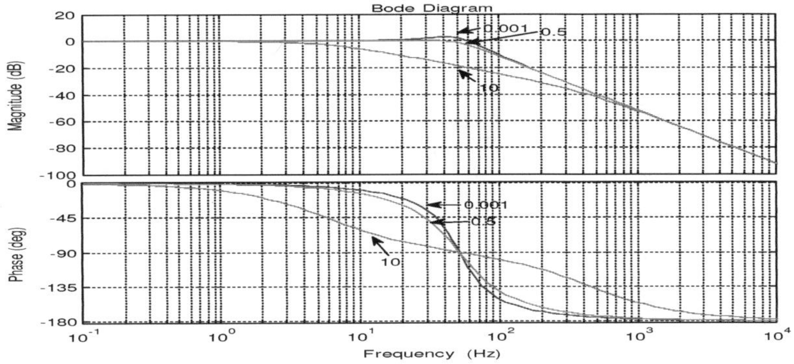

Figure 3. Bode diagram of the sensor with different mutual inductance.

The Figure 3 shows that the proper mutual inductance is also important to the performance of the sensor system. When $M = 0.001$, the mutual inductance of the feedback circuit is too small, and it has little influence on the sensor system. When $M = 0.5$, the sudden change of the amplitude-frequency characteristics can be properly improved. When M is too large, the pass band of the sensor system will be smaller, and it will make the performance of the system worse.

when the parameters of the sensor are determined, the proper matching of mutual inductance and feedback resistance of the feedback circuit of the system can improve the performance of the system. In this design, the feedback resistance Rf=400Ω, the mutual inductance M=1. Thus, the performance of the system can be better.

2.2 Analysis of Impulse Response and Step Response of Magnetic Induction Coil

According to the equivalent circuit diagram of the induction sensor, the following formula can be obtained from the Kirchhoff theorem.

$$\begin{cases} L_s \bullet \dfrac{di(t)}{dt} + \dfrac{1}{C_s} \bullet \int i(t)dt + R_s \bullet i(t) = e(t) \\ U_0(t) = \dfrac{1}{C_s} \bullet \int i(t)dt \end{cases} \quad (2)$$

then the transfer function is written in standard form.

$$H(s) = \frac{\omega_n^2}{s^2 + 2\zeta \bullet \omega_n \bullet s + \omega_n^2} \quad (3)$$

In the formula (3), the $\omega_n = \dfrac{1}{\sqrt{L_s C_s}}$, and the $\zeta = \dfrac{R_s C_s}{2\sqrt{L_s C_s}}$.

When $R_s^2 C_s^2 - 4L_s C_s > 0$, it is $\zeta > 1$, the transfer function has two real poles, it is an over-damped system; when $0 < \zeta < 1$, The transfer function has two complex poles, and it is under-damped system; when $\zeta = 1$, the transfer function had two identical poles, and it is critical damper system[10,11,12].

(1) $0 < \zeta < 1$, Under-damped Systems

Then the transfer function is written as:

$$H(s) = \frac{\omega_n^2}{(s + \zeta \bullet \omega_n + j\omega_d)(s + \zeta \bullet \omega_n - j\omega_d)} \quad (4)$$

In the formula (4), the $\omega_d = \omega_n \bullet \sqrt{1-\zeta^2}$

Order $\omega_d = \omega_n \bullet \sqrt{1-\zeta^2}$, The step response and impulse response are obtained as follows.

$$u(t) = 1 - e^{-\zeta\omega_n t}\left(\cos\omega_d t + \frac{\zeta}{\sqrt{1-\zeta^2}}\sin\omega_d t\right) \tag{5}$$

$$h(t) = \frac{\omega_n}{\sqrt{1-\zeta^2}} e^{-\zeta\omega_n t} \sin\omega_n\sqrt{1-\zeta^2}\,t$$

(2) $\zeta=1$, Critical Damping

$\zeta=1$, the transfer function has two equal poles. Then the transfer function can be expressed as follows.

$$H(s) = \frac{\omega_n^2}{(s+\omega_n)^2} \tag{6}$$

Step response and impulse response are as follows.

$$u(t) = 1 - e^{-\omega_n t}(1+\omega_n t) \tag{7}$$

$$h(t) = \omega_n^2 t e^{-\omega_n t}$$

(3) $\zeta > 1$, Over Damping

The transfer function has two different poles of negative real numbers, the transfer function is expressed as follow:

$$H(s) = \frac{\omega_n^2}{(s+\zeta\bullet\omega_n+\omega_n\bullet\sqrt{\zeta^2-1})(s+\zeta\bullet\omega_n-\omega_n\bullet\sqrt{\zeta^2-1})} \tag{8}$$

The two poles of the transfer function are as follows:

$$s_1 = -\zeta\bullet\omega_n - \omega_n\bullet\sqrt{\zeta^2-1},\ s_2 = -\zeta\bullet\omega_n + \omega_n\bullet\sqrt{\zeta^2-1} \tag{9}$$

Step response and impulse response are as follows.

$$u(t) = 1 + \frac{\omega_n}{2\sqrt{\zeta^2-1}}\left(-\frac{e^{s_1 t}}{s_1} + \frac{e^{s_2 t}}{s_2}\right) \tag{10}$$

$$h(t) = \frac{\omega_n}{2\sqrt{\zeta^2-1}} e^{-\left(\zeta-\sqrt{\zeta^2-1}\right)\omega_n t} - \frac{\omega_n}{2\sqrt{\zeta^2-1}} e^{-\left(\zeta+\sqrt{\zeta^2-1}\right)\omega_n t}$$

(4) Comparison of three cases

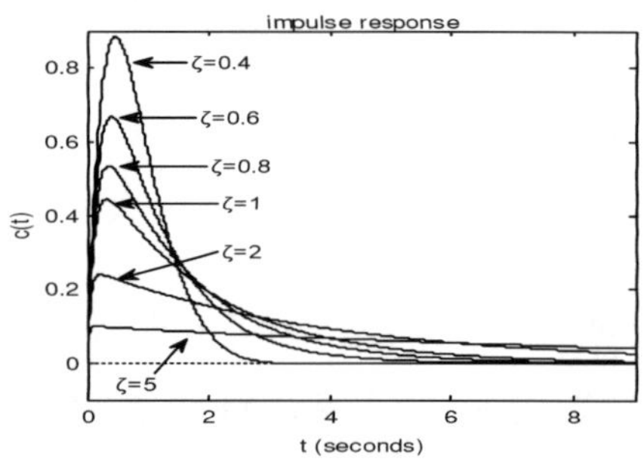

(a)

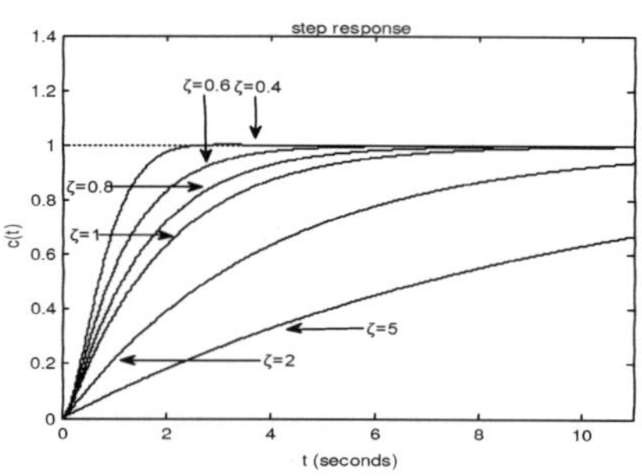

(b)

Figure 4. The impulse response and the step response of the second-order system.

The Figure 4 shows that when $\zeta=0.4, 0.6, 0.8$, the response time is shorter, which avoids the oscillation caused by too small damping ratio and shorten the response time caused by too small damping ratio. It takes about several seconds to reach stable. When the $\zeta \geq 1$, the step response curve rises slowly. the response of the system has no oscillation, but the adjustment time is longer, it takes more than 10 seconds to reach stable in Figure 4 it seriously affects the accuracy of system. Therefore, in the design of the system, the damping ratio should be reasonably determined and it should be controlled in 0.4~0.8.

3. Effect of Feedback Circuit on System Response

When the system adopts negative feedback circuit, the damping ratio of the sensor can be expressed as follows [13].

$$\zeta = \frac{R_s C_s + G \bullet M / Rf}{2\sqrt{L_s C_s}} \tag{11}$$

Then the transfer function of the system is expressed as follows [14].

$$H(j\omega) = \frac{G}{1 - \omega^2 L_s C_s + j\omega(R_s C_s + GM / R\mathrm{f})} \tag{12}$$

3.1 Effect of Feedback Resistance Rf on System Response

The damping ratio is not only related to the parameters of the induction sensor, but also to the mutual inductance coefficient *M* and the feedback resistance *Rf* of the feedback circuit. When the parameters of the induction sensor is determined and the mutual inductance coefficient *M*=1H, the step response and impulse response of the system with different feedback resistors are drawn as follows.

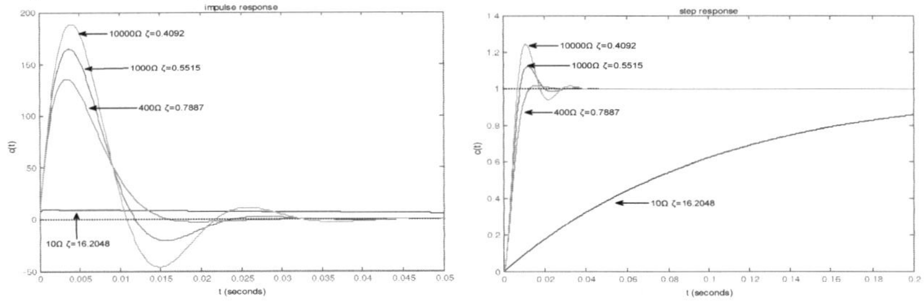

(a) (b)

Figure 5. The impulse response and the step response of the sensor with different feedback resistors.

The Figure 5 shows that the appropriate matching of feedback resistance directly affects the damping ratio. It shows that the system transits from under-damped state to over-damped state with the increase of resistance. When *Rf*=10Ω, the over-damped coefficient reaches 16.2048, which makes the system response time longer and it is difficult to achieve stability; When *Rf*=10000Ω, 1 000Ω and 400Ω, the system is under-damped and tends to be stable after a short period of oscillation. This is consistent with conclusion that the damping ratio should be controlled between 0.4 and 0.8 in the previous analysis. At the same time, when *Rf* = 400Ωand the damping ratio ζ= 0.7887, the system has the best performance.

3.2 Effect of Mutual Inductance on System Response

When the parameters of the magnetic induction sensor have been determined, the feedback resistance *Rf* = 400Ω, and the mutual inductance is different, the impulse response and step response of the system are drawn as follows.

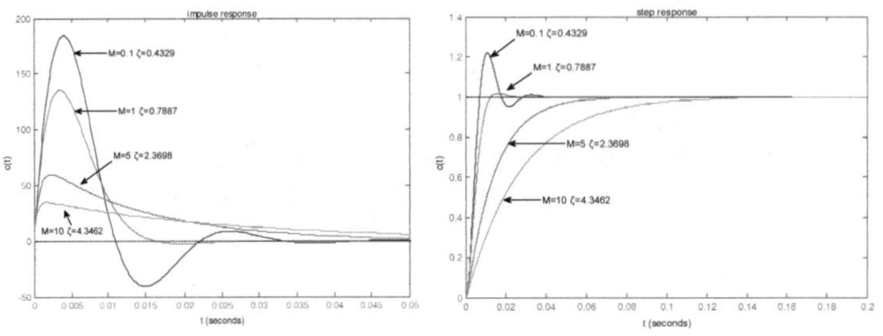

(a) (b)

Figure 6. The impulse response and the step response of the sensor with different mutual inductances.

The Figure 6 shows that the mutual inductance of the feedback circuit also directly affects the damping ratio. With the increase of mutual inductance, the system transfer from under-damped state to over-damped state. When *M*=5H and *M*=10H, the system is in over-damped state, and the response time of the system is longer, so it is difficult to achieve stability. When *M*=0.1H and *M*=1H, the system is under-damped state and tends to be stable after short-time oscillation, this is consistent with conclusion that the damping ratio should be controlled between 0.4 and 0.8 in the previous analysis. At the same time, when M = 1H and damping ratio $\zeta = 0.7887$, the system has the best performance.

In conclusion, when the parameters of the induction sensor are determined, the system performance can be more stable by designing appropriate mutual inductance and feedback resistance to make the damping ratio close to 0.8. In this design, the feedback resistance *Rf*=400Ωand the mutual inductance *M*=1, and the damping ratio $\zeta = 0.7887$.

4. Conclusion

In this article, it analyzes the influence of mutual inductance and feedback resistance in the negative feedback circuit on the system performance in order to design the sensors with better performance. It concludes that if the resistor of the sensor $R = 2.488 \times 10^4 \Omega$,the feedback resistor *Rf*=400Ω, and the mutual inductance *M*=1 to make sure damping ratio ζ be around 0.8, and the sensitivity of the sensor can reach the optimal performance. It can conclude that when the parameters of the induction sensor are determined, the system can achieve the optimal stability performance by matching mutual inductance and feedback resistance reasonably. This is very essential to the stability of the sensor in the design.

References

[1] Cao B, Qin Q M and Li B Sh. Day and night alter-nation of natural ultra-low frequency electromagnetic field and its stability analysis. *Journal of Beijing University: Science and Technology*, 2008, 44(6):897-901.

[2] Wang Y Zh, Cheng D F, Wang J, et al. Research of broad frequency difference magnetic field sensor based on nanocrystalline alloy. *Chinese Journal of Sensors and Actuators*, 2007, 20(9):1967-1970.

[3] He J Sh. Wide field electromagnetic sounding method. *Journal of Central South University: Science and Technology*, 2010,41 (3):1065-1072.

[4] Xu X X. The application of magnetotelluric sounding to the investigation of ore resources at depth. Geophysical & Geo chemical explotation, 2011, 35(1):17-19,23.

[5] Y.Zhao etc. Theoretical modeling of the Aux/IAA negative feedback circuit in plants. *South African Journal of Botany*. 2015, 09:16-19

[6] Guiyuan Wang etc. Stability and hopf bifurcation analysis in a delayed three-node circuit involving interlinked positive and negative feedback loops. *Mathematical Biosciences*. 2019,04.:50-64

[7] David C.Marciano etc. Negative feedback in genetic circuits confers evolutionary resilience and capacitance. *Cell Reports*. 2014.07:1789-1795.

[8] Shao Y Q, Wang Y Zh and Cheng D F, et al. Development of broad frequency magnetic field sensor based on flux feedback. *Chinese Journal of Scientific Instrument*, 2010, 31(11):2461-2466.

[9] Shao Y Q, Cheng D F, Wang Y Zh, et al. Research of high sensitivity inductive magnetic sensor. *Chinese Journal of Scientific Instrument*, 2012, 33(2):349-355.

[10] Wang H J. Characteristics of damping coefficient effect on transient electromagnetic signal. *Chinese Journal of Geophysics*, 2010, 53(2):428-434.

[11] Claude C. Closed loop applied to magnetic measurements in the range of 0.1-50 MHz Review of Scientific Instruments, 2006, 77:1-7.

[12] Ji Y J, Lin J, Yu S B, et al. A study on solution of transient Electromagnetic response during transmitting current turn-off in the ATTEM system. *Chinese J. Geophys (in Chinese)*. 2006, 49(6):1884-1890.

[13] Zang Y L, Xin M W. Quantitative Analysis of Eddy Current Loss in Ferromagnetic Materials. *Journal of Shandong University of Technology (Natural Science Edition),* 2003 (3): 82-85.

[14] Duan Z, Li T B, Li Y S, et al. Application of TEM for geologic prediction at Tongluoshan Tunnel, *Modern Tunneling Technology (in Chinese),*2008,45(2):58-62.

Research on Pose Estimation of Mobile Robot Based on Convolutional Neural Network

Min Yue[1*], Guangyuan Fu[1] and Ming Wu[1]

[1]Xi'an research institute of high-tech, Shanxi, 710025, China

[*]Email: minyue999@126.com

Abstract. In this paper, we propose a new learning scheme for generating camera pose to be used for visual odometry. Pose estimation of mobile robot in unknown environment is formulated as a learning problem. We train a convolutional neural network end to end to compute the 6-dof pose of the robot from a series of images. The architecture is a kind of residual network with multiple attention resblocks, which can give features different weights in order to improve the accuracy of pose prediction. The network estimates ego-motion taking monocular image sequences as input instead of separate images. We call this novel network architecture Posenet. Compared to traditional methods, results are more accurate and in which we can see the potential of pose estimation methods with convolutional neural network.

1. Introduction

Pose estimation is a part of the front-end (visual odometry) in the robot's simultaneous localization and mapping problem [1] which is a necessary prerequisite and preparation for the robot to move autonomously and understand the unknown world. Based on pose estimation, the robot can construct the map of its environment without any sensors other than a camera, thus completing various tasks in many fields. In recent years, SLAM has received more and more attention. There are two main methods to solve SLAM problem of mobile robots, namely filtering method [2-3] and keyframe-based method [4]. In the early SLAM field, the filtering method is mainstream and widely used, but it has some disadvantages such as huge computational cost and tendency to drift. Therefore, recently keyframe-based slam has become more and more popular. We do not need to process every frame of the image sequences so that this kind of method can improve the real-time performance. PTAM [5] is the most famous representative, which divides slam problem into two threads for the first time.

At the same time, some novel methods to solve the pose estimation in SLAM problem have appeared, which are combined with other fields. With the rapid development of deep learning, many fields, such as image recognition and image segmentation [6-8], have made good progress by deep conventional neural network, and the combination of deep learning and SLAM has become one of the research hotspots in the field of robotics. Pose estimation method based on convolutional neural network has attracted more and more attention due to its potential in learning ability and robustness in complex environment.

In this paper, we propose a pose estimation method of monocular robot based on deep learning. The proposed method utilizes residual network to learn the geometric relationship between the image sequences, and the network called Posenet operates in an end-to-end manner, as shown in Figure 1. There are three contributions in this paper: 1) a residual network structure with attention resblock is proposed to improve the accuracy of pose estimation. 2) taking image sequence as input rather than a single image is more conducive for neural network to model features which have temporal characteristics. 3) it proves the application potential of deep learning in robot pose estimation and even slam problem.

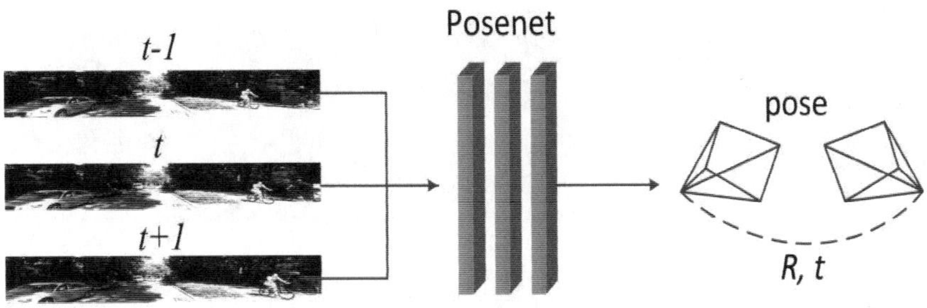

Figure 1. Overview of the proposed Posenet system.

2. Network Architecture

The structure of pose estimation network (Posenet) constructed in this paper is shown in Figure 2. Continuous monocular images are used as input to generate 6-dof pose output. The posenet is a convolutional neural network structure based on ResNet. The original resblock is replaced by the attention resblock to provide weight information for the feature map. It needs three consecutive monocular frames of the image as input in order to predict the pose transformation of intermediate image to the former and the latter frame respectively. Besides, this paper uses rotation and translation errors respectively to construct the loss function and minimize it. The network estimates the 6-dof pose through the end-to-end manner.

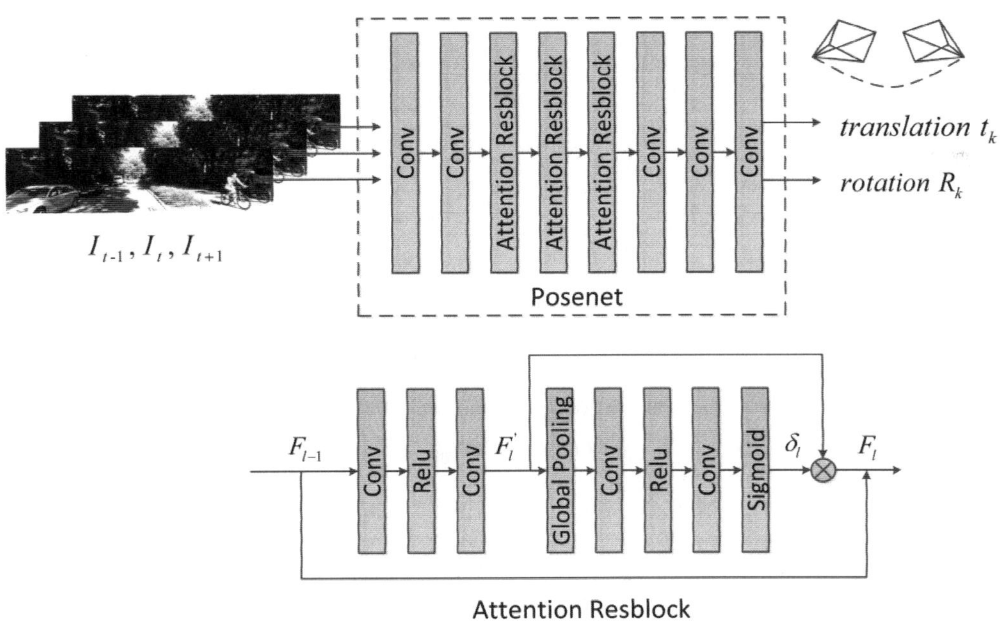

Figure 2. Architecture of the proposed Posenet system.

2.1. Attention Resblock

Different from the traditional resblock in ResNet, the attention resblock can add different weights to the channels in feature map obtained by convolution. Therefore, it is very suitable to deal with problems in which features do not need to be extracted from all pixels of the image or there are some invalid and interfering parts in the image. In the process of pose estimation using convolutional neural network, we take the whole image as the input. However, we can find that there are some pixels or parts of the image that are invalid in pose calculation, such as moving cars and objects that cannot always be photographed in adjacent images as seen in figure 3. These parts interfere with the learning process of the network. Therefore, attention resblock can reduce or eliminate the influence of these parts by adding weights to feature channels.

Figure 3. The invalid or interfering parts in the image. Figure (a) and (b) are the same as figure (b) and (d) and (b) and (d) are the next frames of (a) and (c). The girl on the bike covered by a white square in figure (c) is invisible in figure (b) and (d). Besides, the two people on the bike are both moving which would interfere with the pose estimation.

The structure of attention resblock is shown in Figure 2. In the paper [9] and [10], the authors use it to solve the problem of image super-resolution. In fact, the nature of attention resblock is the CA module (channel attention mechanism) [11]. For some specific problems, the importance of each channel is different. We can increase the weight of important channels and decrease the weight of less important channels through the attention mechanism. Let $F_l(l=1,...,L)$ denote the intermediate feature in the lth attention resblock. The channel attention coefficient δ_l can be calculated as

$$\delta_l = F_1(F_2(F_3(F_l^{'}))),$$
$$F_3(F_{l,c}^{'}) = \frac{1}{h \times w}\sum_{i=1}^{h}\sum_{j=1}^{w} F_{l,c}^{'}(i,j), \tag{1}$$

where F_1 and F_3 denote the sigmoid function for normalization and the global average pooling function respectively, F_2 denotes the combining operation which includes two convolutional layers with a 1×1 kernel size and a ReLu activation function in between, $F_{l,c}^{'}(i,j)$ denotes the value at position (i,j) in the cth channel of the intermediate feature in the lth attention resblock, h and w are the shape of the feature map.

After obtaining the channel attention coefficient δ_l, the output of the lth attention resblock F_l can be calculated as

$$F_l = F_{l-1} + \delta_l \cdot F_l^{'}. \tag{2}$$

2.2. Posenet

The Posenet proposed in this paper is mainly based on ResNet, and its structure is shown in Figure 2. The network takes the original image sequence as input rather than the preprocessed images (such as optical flow images) to learn and predict the camera's pose transformation. Specifically, we need to combine each frame image with its previous and subsequent images as input, as shown in figure 4, that is I_{t-1}, I_t, I_{t+1}, to learn and predict the pose transformation of the camera at time t to time $t-1$ and time $t+1$ respectively which are represented by $T_{t\to t-1}$ and $T_{t\to t+1}$. Using $(h, w, 3)$ to represent the shape of a single image, the shape of the input image is $(h, 3\times w, 3)$. The network has 5 convolutional layers and 3 resblocks, and two 6-dof pose can be obtained after the last convolution.

Figure 4. The input image sequence.

3. Training Procedure

3.1. Loss functions

The pose of robot can be divided into rotation and translation. We calculate the errors of these two parts respectively and the final loss function L_{motion} for the motion vectors is

$$L_{motion} = loss(T_{t\to t-1}) + loss(T_{t\to t+1}),$$
$$loss(T) = \left\| R_{gt} - R_{pred} \right\|_2 + \left\| t_{gt} - t_{pred} \right\|_2 . \tag{3}$$

In equation (3), R_{gt} 和 t_{gt} represent the angle of the rotation and the normalized translation ground truth which matches $\|t\|_2 = 1$, R_{pred} 和 t_{pred} denote the 6-dof prediction of motion respectively. These four variables all have three parameters, R_{gt} and R_{pred} both corresponding to $[r_x, r_y, r_z]$ and t_{gt} and t_{pred} both corresponding to $[t_x, t_y, t_z]$.

3.2. Training Schedule

The network training is based on the tensorflow framework with Ubuntu 16.04 system. We train the model from scratch with Adam [12] using a momentum of 0.9 and a weight decay of 0.0002. All of the experiments are based on monocular image sequences. In order to adapt to the requirements of the convolutional neural network for the input image size, this paper carries out a unified operation on all the images in the training process and adjusts the image size. So the shape of the combined three image frames input is $128\times 1248\times 3$, and different kinds of data enhancement methods are used to improve performance and reduce possible overfitting, such as selecting images for scaling randomly. Finally, the posenet is trained for 300k iterations with all other weights fixed.

4. Experiments

4.1. Datasets

We use KITTI dataset as the training and evaluation dataset which is the most common benchmark used in prior work for evaluating the accuracy of pose prediction. The KITTI dataset includes a full suite of data sources such as stereo video, 3D point clouds from LIDAR, and the vehicle trajectory.

Kitti contains real image data from urban, rural and highway scenes, with up to 15 cars and 30 pedestrians in each image, as well as varying degrees of occlusion and truncation. We select the images of camera 02 in KITTI odometry dataset and the first eight sequences as the training set and the last two as the validation set. The number and size of image frames are shown in Table 1.

Table 1. KITTI odometry benchmark sequence 00-10.

Number	Sequence name	Image numbers	Image size
00	2011_10_03_drive_0027	4541	1241,376
01	2011_10_03_drive_0042	1101	1241,376
02	2011_10_03_drive_0034	4661	1241,376
03	2011_09_26_drive_0067	801	1241,376
04	2011_09_30_drive_0016	271	1226,370
05	2011_09_30_drive_0018	2761	1226,370
06	2011_09_30_drive_0020	1101	1226,370
07	2011_09_30_drive_0027	1101	1226,370
08	2011_09_30_drive_0028	4071	1226,370
09	2011_09_30_drive_0033	1591	1226,370
10	2011_09_30_drive_0034	1201	1226,370

4.2. Evaluation of Pose Estimaiton

After training the Posenet, the ego-motion of robot is learned from monocular images. Table 2 reports the ego-motion accuracy of our modes over two sample sequences from the KITTI odometry dataset. The evaluation index is absolute trajectory error(ATE). We can find out that our proposed method outperforms the traditional methods of ORB-SLAM [13] (short).

Table 2. Absolute Trajectory Error (ATE) on the KITTI odometry dataset.

Method	Seq. 09	Seq. 10
ORB-SLAM(full)	0.014+0.008	0.012+0.011
ORB-SLAM(short)	0.064+0.141	0.064+0.130
Ours	0.023+0.027	0.021+0.016

5. Conclusions and Future Work

In this paper, we propose the Posenet, a pose estimation method for mobile robot based on deep learning, and attention resblock is used to distinguish the weights of features extracted from images. The input of the network is monocular original image sequence with three consecutive frames instead of a single or preprocessed image. The proposed method performs better and has more potential than the traditional structure from motion. The next step is focus on improving the robustness and training efficiency of the network.

References

[1] Liu H, Zhang G and Bao H 2016 A survey of monocular simultaneous localization and mapping *J. Comp-Aid. Des. Comp. Graph.* **28** 855-868

[2] Davison A J, Reid I D, Molton N D and Stasse O 2007 Monoslam: real-time single camera slam *IEEE Trans. Pat. Anal. Mach. Intel.* **29(6)** 1052- 1067

[3] Montiel J M M, Civera J and Davison A J 2006 Unified inverse depth parametrization for monocular slam *IEEE. Trans. Robot.* 10.15607/RSS.2006.II.011

[4] Mouragnon E, Dekeyser F, Sayd P, Lhuillier M and Dhome M 2006 Real Time Localization and 3D Reconstruction *Proc. IEEE. Comp. Soc. Conf. Comp. Vis. Pat. Rec.* **1** 363- 370

[5] Klein G 2007 Parallel tracking and mapping for small AR workspaces *Proc. Sixth. IEEE. ACM. Int. Sym. Mix. Aug. Real.* 225-234

[6] Nekrasov Vladimir, Shen Chunhua and Reid Ian 2018 Light-Weight RefineNet for Real-time semantic segmentation

[7] Chen L C, Papandreou G, Kokkinos I, Murphy K and Yuille A L 2016 Deeplab: semantic image segmentation with deep convolutional nets, atrous convolution, and fully connected crfs *IEEE. Trans. Pat. Anal. Mach. Intel.* **40(4)** 834-848.

[8] Ren S, He K, Girshick R and Sun J. 2015 Faster r-cnn: towards real-time object detection with region proposal networks *IEEE. Trans. Pat. Anal. Mach. Intel.* **39(6)** 1137-1149.

[9] Zhang Y, Li K, Li K, Wang L, Zhong B and Fu Y 2018 Image super-resolution using very deep residual channel attention networks

[10] Shi Z, Chen C, Xiong Z, Liu D, Zha Z J and Wu F 2018 Deep Residual Attention Network for Spectral Image Super-Resolution *Eur. Conf. Comp. Vis. Spr. Cham.*

[11] Hu J Shen L, Albanie S, Sun G and Wu E 2017 Squeeze-and-excitation networks Spectral Image Super-Resolution *Eur. Conf. Comp. Vis. Spr. Cham.*

[12] Kingma D P and Ba J 2014 Adam: a method for stochastic optimization *Int. Conf. Learning. Representations.*

[13] Mur-Artal R, Montiel J M M and Tardos J D 2015 ORB-SLAM: a versatile and accurate monocular slam system *IEEE. Trans. Robot.* **31(5)** 1147-1163

IOP Conf. Series: Materials Science and Engineering **565** (2019) 012014 doi:10.1088/1757-899X/565/1/012014

Influence of Waveform Symmetry on Output Performance of Piezoelectric Inertia Actuator Controlled by Composite Method

H Y, Qi*

Research Institute of Automatic Control Engineering, NFC Metallurgical Machinery Co., Ltd. Shenyang, No. 26, Xiheshi North Street 110027, China

*Email: qi-hong-yan@163.com

Abstract. The waveform symmetry has a great influence on the piezoelectric inertia actuator. In this paper, the waveform symmetry based on the composite method is explained and its influence on the output performance is discussed in relation to four aspects: frequency characteristics, voltage characteristics, load capacity, driving capacity. The designed prototype is fabricated and a series of experiments are carried on. Experiments show the waveform symmetry based on the composite method is widened effectively relative to the traditional driving method. The high-speed of the actuator is obtained under 60% and 40%.The large-load is achieved under 90% and 10%. When the saw-tooth driving waveform voltage is 30 V_{p-p} for 800 Hz and a sinusoidal friction regulation waveform voltage is 6 V_{p-p} for 39 kHz, the maximum velocities in the forward and reverse direction are 1.48 mm/s and 1.39 mm/s. Meanwhile, the maximum vertical loads of the actuator are 540 g and 510 g. The actuator driven by different symmetrical composite waveforms will make it ideal for miniature devices.

1. Introduction

Piezoelectric inertia actuator has been developed for several decades because of their compact structure, immunity from magnetic interference, theoretically unlimited movement, low production cost [1-4]. Given these advantages, the inertia actuator have become viable candidates for using in camera focusing mechanisms, cell phones, scanning probe microscopes, zoom lens systems, and blue-ray devices. The specifications of inertia actuators mainly contain velocity, load, and driving capacity. The tradeoff between the velocity and load should be considered depending on different applications. The driving capacity is generally expected to be high for fitting into the miniature devices [5-6].

As one type of driving methods for piezoelectric inertia actuators, a traditional saw-tooth waveform is generally used as an exciting signal that includes slow and rapid deformation stages. However, the kinetic friction force in the rapid deformation stage frequently results in backward motion [7-10]. The composite method can restrained the backward motion effectively. It need to be pointed out that the waveform symmetry of the composite method has a great influence on output performance of inertia actuator. Thus, the goal of this work is to provide a comprehensive research on the symmetry of the proposed method. Previously, researchers have done some work related to waveform symmetry on performance aspect of actuator. Yuting Ma *et al.* researched the performance of the resonant-type inertial impact motor driven by the different symmetric rectangular pulse methodology. With the optimized waveform symmetry, both the velocity and load could be enhanced. Thus, there is an

IOP Conf. Series: Materials Science and Engineering **565** (2019) 012014 doi:10.1088/1757-899X/565/1/012014

optimal symmetry value to generate the high-speed and large-load capacity, which could be used in practical applications [11]. Chia-Feng Yang *et al.* studied the stepping motion of the piezoelectric device based on various symmetric triangular waveforms. This work proposed an impulse model to describe its motion behavior and indicates that not all inputs of waveform symmetry can work well. A proper symmetry can make piezoelectric actuator more suitable for use in high precision position and control [12]. Yuxin Peng *et al.* presented that the waveform symmetry has a distinct effect on the driving characteristics of linear actuator, especially for the precision positioning of the dual objects. The results indicate that the proposed actuator excited by different waveform symmetry can be better applied to different positioning applications [13]. Although there have been some studies on the waveform symmetry, it is still necessary to research the different symmetric composite method on the performance of inertia actuator. This paper will further provide an important guiding principle for the proposed method in practical applications.

2. Waveform symmetry

Figure 1 illustrates waveform symmetry distribution diagram of composite method in the forward and reverse direction. This method is realized by a composite waveform that includes a saw-tooth driving waveform and a sinusoidal friction regulation waveform, and the sinusoidal friction regulation waveform is applied to the rapid deformation stage of the saw-tooth driving waveform.

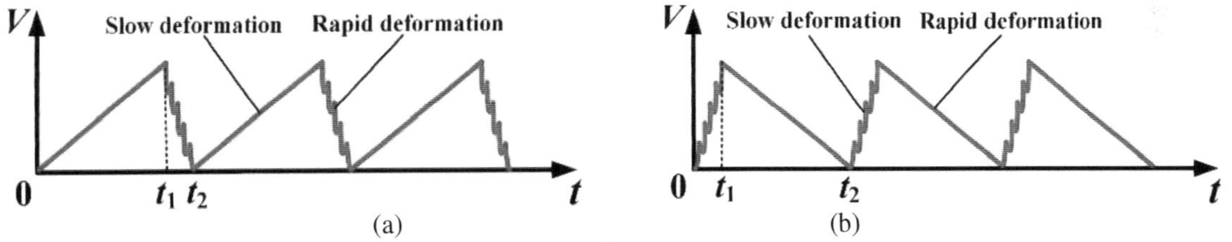

Figure 1. Waveform symmetry distribution of composite method. (a) is the waveform symmetry distribution in forward direction; (b) is the waveform symmetry distribution in reverse direction.

To describe the performance of actuator by different waveform symmetry, the waveform symmetry S based on composite method is defined as following:

$$S = \frac{t_1}{t_2} \times 100\% \tag{1}$$

where S is waveform symmetry of composite method; 0~t_1 is the time of driving voltage going up; 0~t_2 is driving cycle. The waveform symmetry is changed by adjusting time ratio of t_1 and t_2. Figure 1(a) and (b) show that the hybrid waveform in the forward and reverse direction.

3. Driving principle

For an inertia actuator, an asymmetric motion is essential to provide stick and slip motions with its friction partners. The composite waveform seen in Figure. 1 is selected to drive the inertia actuator. According to Figure 2, the operation process of actuator in a driving cycle is explained in detail. When the applied voltage increases at relatively low speed, the slider moves with the stator along to the x direction; when the voltage decrease quickly, the stator goes back to initial position and the slider slips against the stator in the x direction due to inertial force. The backward motion is generated due to kinetic friction force between the slider and stator. Compared with traditional sawtooth waveform, the composite waveform can excite the stator to generate a micro vibration in rapid deformation stage. By this micro vibration, the backward motion of the actuator is obviously restrained. The large effective step displacement ΔDS in one direction is obtained, and a continuous stepping motion is achieved by cyclic composite waveform. By changing the waveform symmetry, the moving direction of the slider can be reversed.

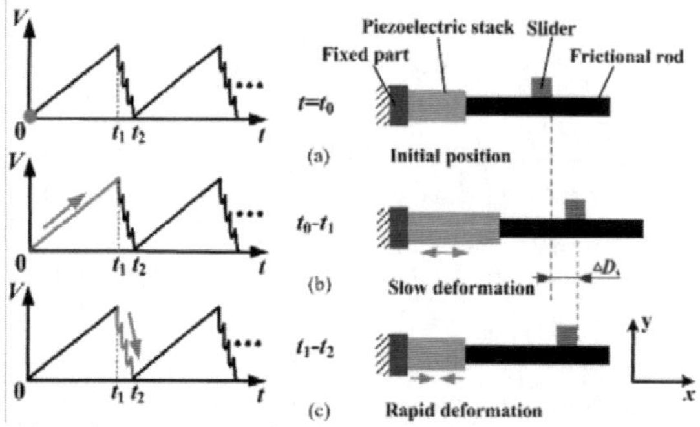

Figure 2. Driving principle of inertia actuator under the composite waveform. (a) is the initial position;(b) is the slow deformation stage;(c) is the rapid deformation stage.

4. Experimental system

Figure 3 is the established tested system. The main parts of the system includes a computer, a power analyzer (NORMA4000, Fluke Co. Ltd, America), a laser sensor (LK-H020, Keyence Co. Ltd, Japan), an arbitrary waveform generator (DG4162, Beijing RIGOL Technology Co. Ltd, China), a power amplifier (XE500-C, Harbin Core Tomorrow Science & Technology Co. Ltd, China), and a actuator. When the tested system works, the arbitrary waveform generator controlled by the computer is used to provide a composite waveform. The arbitrary waveform generator voltage range is from -10 V to +10 V. The voltage is amplified by power amplifier, and the enlarged voltage signal is used to drive the prototype. The prototype is excited to generate a continuous output motion. A laser sensor with the resolution of 20 nm is used to measure its motion behavior, and the test data are gathered by its built-in software via computer. Meanwhile, the power analyzer is used to measure its input power.

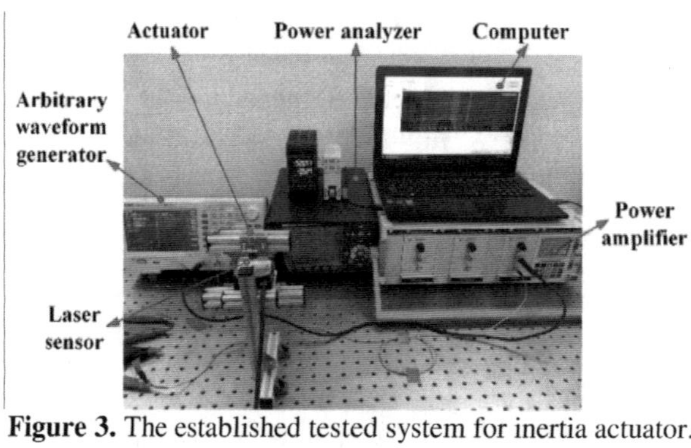

Figure 3. The established tested system for inertia actuator.

5. Results and discussion

5.1. Symmetry characteristics

The parameters of the traditional sawtooth waveform are the same as the saw-tooth driving waveform. For the traditional driving method, the designed actuator cannot work normally in between 30%~50% and 50%~70% in both directions because no obvious saw-tooth type displacement is output. The large velocity can be obtained when the waveform symmetries are near 90% and 10%. The waveform symmetry in between 75%~90% and 10%~25% can be used in practical applications, as shown in figure 4 (a). For the composite method, the high velocity can be easily obtained relative to traditional sawtooth waveform. When the waveform symmetry is 50%, the velocities of the actuator can reach 1.72 mm/s and 1.74 mm/s in forward and reverse direction, as shown in figure 4 (b). The waveform diagram of 50% in forward and reverse direction is different. The larger operation range of the composite method is realized. Thus, the waveform symmetry is widened effectively relative to traditional driving method.

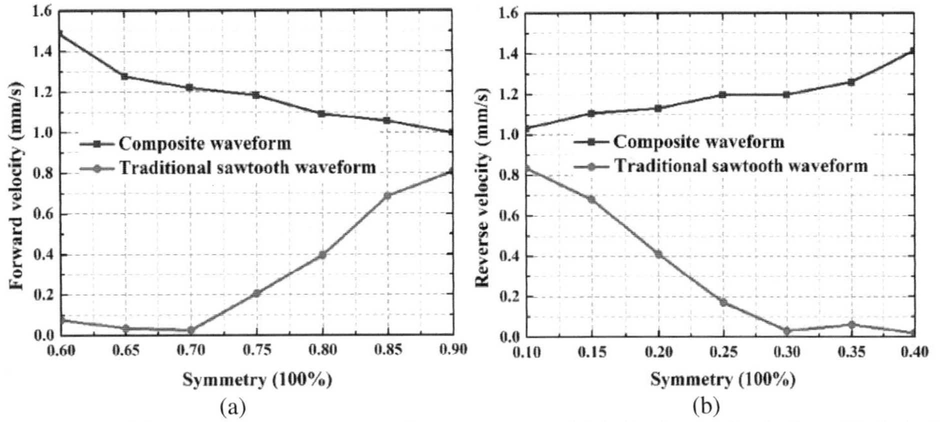

Figure 4. Velocity of the actuator versus to waveform symmetry. (a) is the forward velocity with the hybrid and traditional driving waveform; (b) is the reverse velocity with the hybrid and traditional driving waveform.

5.2. Frequency characteristics

Figure 5 shows the curve of the output velocity versus the driving frequency with different waveform symmetry at the load of 30 g when the saw-tooth driving waveform voltage is 30 V_{p-p}, and the sinusoidal friction regulation waveform voltage and frequency are 6 V_{p-p} and 39 kHz. The velocity of the actuator has an increasing trend with the frequency from 0 Hz to 800 Hz. The results indicate that the velocity of the actuator under different waveform symmetry has a slight change in between 300 Hz and 500 Hz in both directions .Before the frequency of 300 Hz, the high velocity is realized under 90% and 10%, and the actuator starts to work at 50 Hz. The larger frequency operation range can be achieved due to the small driving frequency, which is more suitable for velocity control of fine motion. After the frequency of 300 Hz, the actuator under the symmetries of 60% and 40% can achieve high velocity, and the output velocities are 1.47 mm/s and 1.46 mm/s at 800 Hz in two directions. The relationship between the frequency and velocity is not linear. This may be caused by preloading gap and assembly errors.

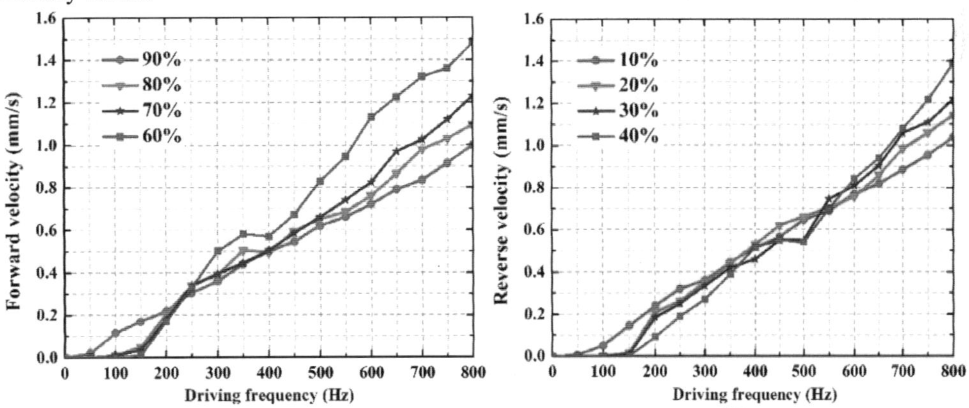

Figure 5. Velocity versus to frequency at the vertical load of 30 g. (a) is the forward velocity with symmetries of 60%~90%; (b) is the reverse velocity with symmetries of 10%~40%.

5.3. Voltage characteristics

Figure 6 shows the relationship between velocity and driving voltage in both directions at 30 g when the saw-tooth driving waveform frequency is 800 Hz, and the sinusoidal friction regulation waveform voltage is 6 V_{p-p} for 39 kHz. The velocity linearly increases with driving voltage from 5 V_{p-p} to 60 V_{p-p}. The results indicate that there is no significant change in the performance of the actuator under different waveform symmetries before the voltage of 20 V_{p-p}, and the minimum driving voltage is 5

$V_{p\text{-}p}$. Thus, the actuator is particularly applied to high precision position and control, and its work is not easy to generate the heat. After the voltage of 20 $V_{p\text{-}p}$, the waveform symmetry has an obvious influence on the actuator. When the voltage is 60 $V_{p\text{-}p}$ the actuator can achieve the velocities of 3.48 mm/s and 3.26 mm/s under 60% and 40%. Meanwhile, under 90% and 10% the velocities of the 2.27 mm/s and 2.29 mm/s can be obtained. The actuator is suitable for zoom lens systems and camera lens modules that require high velocity.

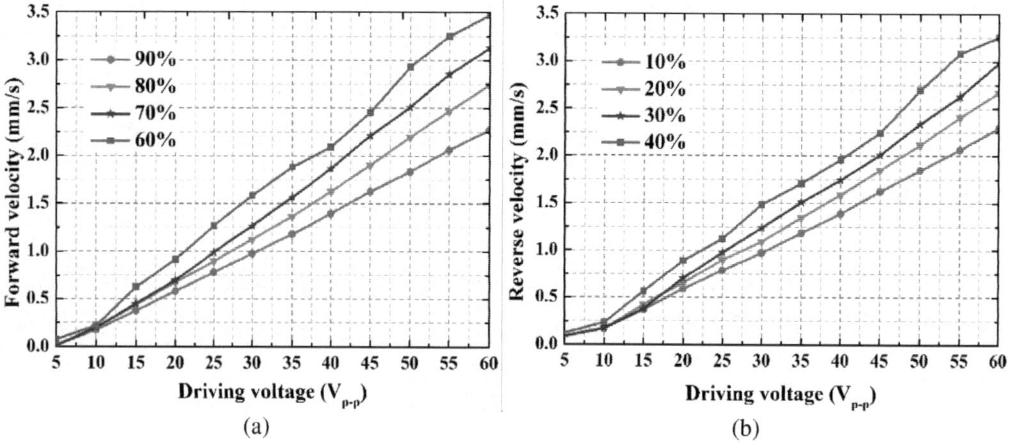

Figure 6. Velocity versus to voltage at a vertical load of 30 g. (a) is the forward velocity with the symmetries of 60%~90%; (b) is the reverse velocity with the symmetries of 10%~40%.

5.4. Load capacity

The standard weight is used to investigate load characteristics along negative direction. The relationship between vertical load and velocity is illustrated in Figure. 7. The initial mass of slider is 30 g without load. The various loads were loaded on the slider during the experiments. It is seen that the velocity decreases when the vertical load becomes large. The forward maximum vertical loads are 540 g when the symmetries are 90%. The reverse maximum vertical loads are 510 g when the symmetries are 10%. Thus, the actuator excited by the hybrid driving waveform could be suitable for the information technology devices that require large-load operation. In addition, the velocity of the actuator are nearly equivalent in both directions when the vertical load is 210 g, the large velocity can be obtained under the symmetries of 90% and 10%. Thus, a larger stiffness of the actuator is obtained relative to the symmetries of 90% and 10%.

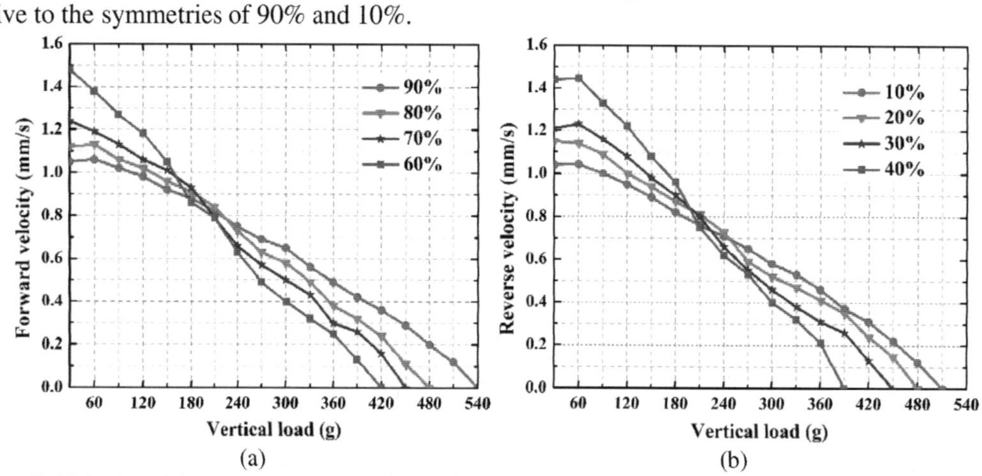

Figure 7. Velocity of the actuator versus to the vertical load. (a) is the forward velocity with the symmetries of 60%~90%;(b) is the reverse velocity with the symmetries of 10%~40%.

Taking into account the potential application value of the actuator, the load capacity excited by different symmetrical hybrid actuation waveform is further quantified. An important of the mass ratio λ is defined by as follows:

$$\lambda = \frac{M_{\mathrm{m}}}{M_s} \tag{2}$$

where λ is the mass ratio, M_{m} is the maximum vertical load, M_{s} is the mass of stator, and its mass is 4 g.

The maximum vertical load and mass ratio are listed in Table 1.

Table 1. The maximum vertical load and mass ratio in forward direction and reverse direction.

Symmetry	Forward direction				Forward direction			
(100%)	90%	80%	70%	60%	40%	30%	20%	10%
M_{m} (g)	540	480	450	420	390	450	480	510
M_{s} (g)	4	4	4	4	4	4	4	4
λ	135:1	120:1	113:1	105:1	98:1	113:1	120:1	128:1

According to Table I, the forward mass ratio can reach 135:1 under waveform symmetries of 90%. The reverse mass ratio is 128:1 under symmetries of 10%. Thus, the actuator under 90% and 10% can easily achieve the large-load operation.

6. Conclusions

In this paper, the waveform symmetry based on the composite method is researched. The waveform symmetry is explained, and the driving principle of composite waveform is analyzed. A prototype of the inertia actuator is fabricated and its output characteristics are tested. The tested results indicate that the maximum velocities are 1.48 mm/s and 1.39 mm/s under 60% and 40% when the saw-tooth driving waveform voltage of 30 $V_{p\text{-}p}$ for 800 Hz and sinusoidal friction regulation waveform voltage of 6 $V_{p\text{-}p}$ for 39 kHz. The inertia actuator can be used in practical application that required high-speed operation. The maximum vertical loads of the actuator are 540 g and 510 g under 90% and 10%. Correspondingly, the maximum mass ratio is 135:1 and 128:1. Thus, the actuator can achieve the large-load operation. In addition, the waveform symmetry of the composite method is widened relative to the traditional sawtooth waveform. This paper will provide an important guiding principle for the method in practical applications. Because of the simple structure and low cost of both mechanical and electrical parts, the proposed actuator is a promising information technology devices for use in camera, manipulator systems and blue-ray equipment.

References

[1] Rakotondrabe M, Haddab Y and Lutz P 2008 IEEE T. Contr. Syst. T. **16** 1316.
[2] Meyer C, Sqalli O, Lorenz H and Karrai K 2005 Rev. Sci. Instrum. **76** 063706.
[3] Li J, Zhou X, Zhao H, Shao M, Fan Z and Liu H 2015 Rev. Sci. Instrum. **86** 045002.
[4] Morita T, Yoshida R, Okamoto Y, Higuchi T 2002 Precis. Eng. **26** 289.
[5] Nishimura T, Hosaka H, Morita T 2012 Ultrasonics. **52** 75.
[6] Morita T, Nishimura T, Yoshida R and Hosaka H 2013 Jpn J Appl Phys. **52** 07HE05.
[7] Zhang Q, Chen X, Yang Q and Zhang W 2012 Int. J. Adv. Manuf. Technol. **61** 1029.
[8] Huang H, Zhao H, Yang Z, Mi J, Fan Z, Wan S, Shi C and Ma Z 2012 Rev. Sci. Instrum. 83 055002.
[9] Peng J and Chen X 2011 IEEE-ASME T. Mech. **16** 394.
[10] Wang S, Zhang Z, Ren L. Zhao H, Liang Y, Zhu B 2015 Microsyst. Technol. **21** 1757.
[11] Ma Y, Shekhani H, Yan X, Choi M and Uchino K 2016 Sensor. Actuat. A-Phys. **248** 29.
[12] Yang C, Jeng S and Chieng W 2011 Sensor. Actuat. A-Phys. **166** 66.
[13] Peng Y, Cao J, Guo Z and Yu H 2015 Smart Mater. Struct. **24** 125039.

IOP Conf. Series: Materials Science and Engineering **565** (2019) 012015 doi:10.1088/1757-899X/565/1/012015

The Design of an Intelligent Fast Continuous Vialog

Zhang Jinning[1], Zhou Yishu[1*], GuoHongbo[1] and He Huayang[1]

[1]Research Institute of Highway Ministry of Transport No. 8 Xitucheng Road, Haidian District, Beijing

[*] E-mail: ys.zhou@rioh.cn

Abstract: Surface evenness affects vehicle safety, service life and passenger comfort. A road surface with poor evenness will cause resistance to the running of the vehicle, causing the vehicle to generate vibration bumps, hinder the driving safety of the vehicle, and cause localized water accumulation and accelerated rode damage speed. In order to make the vehicle safe, passengers feel comfortable, road life growth, road flatness detection is particularly important. The detection, evaluation and analysis of the surface evenness, pavement cracks and rutting size of the road surface that has been opened to traffic, it is great significance for the scientific assessment of road construction grades and road surface conditions. In order to overcome the shortcomings of the existing continuous vialog, this paper designed a new type of continuous vialog based on the fast continuous surface evenness algorithm.In this paper, the contact distance measuring device and the measurement method of the ordinary gyroscope are used to realize the sampling, calculation and analysis processing of the elevation value and the surface evenness standard deviation. The experiment proves that the continuous vialog has the advantages of high precision, small error and low cost.

1. Introduction

In recent years, with the rapid development of China's transportation industry, especially the mode of road transportation has played an important role in the vigorous development of the national economy [1].The index of surface evenness directly affects the service life of the road, the driving safety of the vehicle and the comfort level of the human body [2].Therefore, in order to make the vehicle safe to travel, the passenger feels comfortable, the road life is increased, and the detection of the road surface evenness is particularly important.Currently, the commonly used surface evenness evaluation indexes include: ARS average adjustment slope, BI bumpy accumulation index, CP surface evenness index, QI 1/4 vehicle index, NAASRA index, international surface evenness index and so on [3].According to the research literature, the algorithm of most continuous vialogs at this stage is to calculate the IRI first, and then calculate the σ value by σ=0.6IRI [4]. This algorithm is derived from Wei Xiaodan in the study of the international surface evenness index IRI as the evaluation index of surface evenness. By using the level gauge and the hand-inferred surface meter, the international surface evenness index IRI value and the continuous surface evenness standard deviation σ are calculated. The value is then tested for correlation between the two indicators. The test results show that the international surface evenness index IRI and the surface evenness standard deviation σ can be converted by the formula σ=0.6IRI.However, this algorithm is calculated by statistics [5]. Theoretically, there is no correlation between the two indexes in mathematics, and different literatures have different conversion methods for the two indexes [6].Among them, Xu Yongxin analyzed the use method and working principle of the surface evenness test vehicle in the "Research on the method of surface evenness detection", and established the corresponding relationship between the surface evenness and the test data, and gave different conversion methods: IRI = $1.0319\sigma + 0.2711$.Based on this, it is found that different devices use IRI value to convertσ, there will be different correlation, simple correlation conversion, the value can not guarantee the requirements, and the algorithm based on IRI value to obtain σ is not universal [7]. Therefore Zhang Jinning et al. proposed a fast continuous continuous vialog algorithm for calculating σ from elevation values. In this paper, a fast continuous continuous vialog is designed based on this surface evenness algorithm to measure the main index elevation value of the algorithm and calculate the standard deviation σ of the surface evenness.

2. Method of measuring elevation values

The biggest difficulty in the design of the fast continuous vialog is to determine the standard elevation value.There are three methods commonly used in the detection of the standard elevation value, namely the leveling method, the laser ranging plus inertial device compensation method, and the gyroscope/angle sensor relative elevation method. The leveling method is to use a level or total station to measure the road surface elevation.This method is a conventional measurement method, because the measurement speed of this method is slow, cannot achieve the purpose of fast, this method is not used.Compensation method based on laser ranging and inertial device generally includes height of non-contact sensors (such as laser, infrared, visible, ultrasonic sensors) accelerometer, and the longitudinal distance, the height of non-contact sensor were measured the vertical distance to the road surface, vertical acceleration and longitudinal distance along the cross section, a section can be obtained in the corresponding calculation formula of elevation data, its working process for the first through laser sensor measuring equipment to the vertical elevation of the road, and then use the acceleration sensor to measurethe absolute displacement of the laser sensor in the vertical direction due to the bump during the running of the vehicle, thereby correcting the laser sensor. In the measurement process, due to the measurement error caused by bumps, the real road surface longitudinal elevation information is obtained.The precision of such devices mainly depends on the laser and acceleration sensor, which are not suitable for this study due to their high price.The gyro/angle sensor relative height method passes the mechanical mechanism to ensure precise control of the step size, and then the angle change is calculated by the angle measuring sensor or the high-precision gyroscope and converted into elevation, and the elevation series is generated by the cumulative method. This method stems from Face's development of an instrument that can replace the leveling method, the Dipstick, which is equipped with a small counter that automatically records elevation data. Although the gyro/angle sensor relative elevation method has better precision and higher efficiency than the level method, it still cannot meet the fast demand. At the same time, such equipment has the highest price in the same kind of equipment, so this solution is not adopted.

Based on this, this paper proposes a relative elevation method using ranging and compensation to measure the elevation value of the road surface, which is completed by using a contact distance measuring device plus an ordinary gyroscope. This method combines the advantages of laser ranging plus inertial device compensation and gyroscope/angle sensor relative elevation. The non-contact measurement in the laser ranging plus inertial device compensation method becomes contact measurement, and the form of the laser is not used, which not only reduces the cost but also increases the accuracy. In addition, the replacement of high-precision gyroscopes and angle measuring devices by ordinary gyroscopes helps to reduce the cost of the equipment.

3. Hardware design

The fast continuous vialog mainly consists of two parts: an elevation measuring device and a gyroscope correction device. The working principle is to fix the rapid continuous vialog on the car, the trailer or the trolley through the bracket, measure the fluctuation of the ground by the elevation measuring device, and at the same time, use the gyroscope to correct the change of the angle of the device and correct it. The mechanical structure of the fast continuous vialog is shown in Figure 1.

IOP Conf. Series: Materials Science and Engineering **565** (2019) 012015 doi:10.1088/1757-899X/565/1/012015

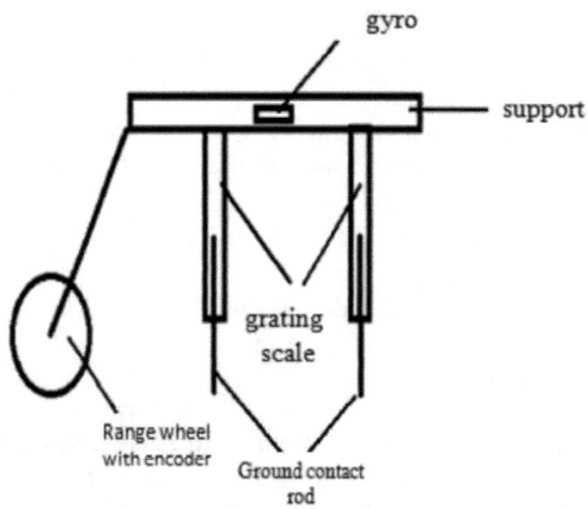

Figure 1. Mechanical structure of fast continuous vialog.

The actual measured road surface elevation value is calculated according to the theoretical formula (1):

$$h_{i+1} = h_i + l_2 - l_1 - l \times sin\theta \quad (1)$$

Where: h_{i+1} is the elevation of the new measuring point;h_i-- the elevation of the original measurement point;l_2-- the indication value of elevation measurement sensor 2;l_1-- the indication value of elevation measurement sensor 1; l-- the distance between two elevation measurement sensors.

The contact type distance measuring device in the elevation measuring equipment uses a grating scale as the measuring element.Grating scale has the advantages of high precision, good stability, fast speed and strong anti-interference ability, and is widely used in high-precision displacement measurement fields such as measuring instruments and numerical control machine tools. The gyroscope correction device adopts the nine-axis MPU9250 accelerometer electronic gyroscope, which has the advantages of small size, light weight, low cost, high reliability, large range and easy intelligence. It is suitable for installation space and weight requirements and overload.

The continuous vialog is mainly composed of a main bracket, an elevation measuring device, a ground connecting rod and a measuring wheel. The control system consists of a scale 1, a scale 2, a grating data acquisition card, an RS232 to TTL level module, a gyroscope, an encoder, an STM32 controller and a touch screen. The system block diagram is shown in Figure 2.

The main bracket, the elevation measuring device and the measuring wheel with encoder are mechanically connected. The gyroscope is embedded on the main bracket to measure the inclination of the device, and the dip angle data is sent to the STM32 through the IIC/SPI protocol; the main structure of the elevation measuring device is the grating scale 1 and the grating scale 2, and the grating scale sliding block is connected to the ground connecting rod and the ground. The connecting rod is close to the ground, and the data collected by the two grating scales is sent to the RS232 to TTL converter according to a specific protocol through the data acquisition card, and converted into a TTL level input to the serial port of the STM32, and the data is parsed by the processor. Obtaining the elevation value of the ground; the measuring wheel with the encoder is in contact with the ground. Through the rotation of the wheel, the encoder coupled with the wheel transmits the level-coded data to the STM32 for encoding and counting, thereby obtaining the moving distance of the device.

STM32 collects the data of the grating scale, gyroscope and encoder, analyzes the data, displays the data on the touch screen, and can control the system configuration through the touch screen.

IOP Conf. Series: Materials Science and Engineering **565** (2019) 012015 doi:10.1088/1757-899X/565/1/012015

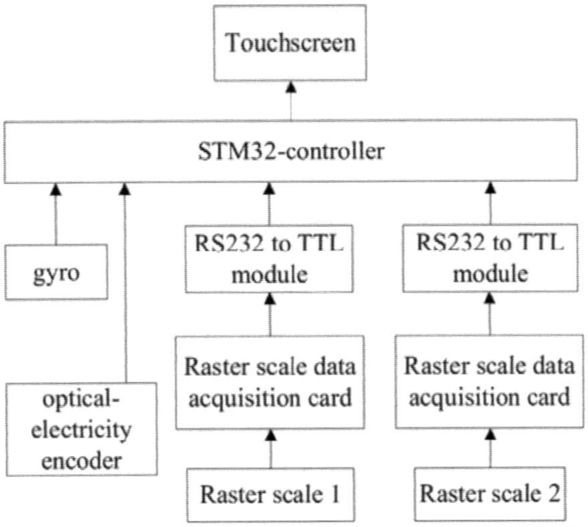

Figure 2. Flow chart of the rapid continuous vialog control system.

The specific test steps are as follows: (1) Before using the device to measure the elevation value of the road surface, the equipment is horizontally calibrated by the gyroscope inclination data to find an absolutely flat road surface. (2) Adjust the two-way ground connecting rod and the encoder measuring wheel fixed by the grating scale to ensure that the two parts fully touch the ground. (3) Start the device to enter the measurement mode. (4) The device performs real-time elevation value detection and stores the elevation value to the SD card. (5) The device detects the moving distance through the encoder, displays and stores it to the SD card. (6) The device detects the tilt angle through the gyroscope and stores it on the SD card. (7) Each value can be read through the system screen, or the height, movement distance and inclination data in the SD card can be read by connecting the USB Slave interface of the STM32 through a PC.

4. Software design

The control system software is programmed by Keil, according to the flow chart shown in figure 3, the system initialization is first carried out, including contact type distance measuring device initialization, gyroscope correction device initialization and encoder initialization; then the grating scale, gyroscope and encoder are sequentially Send instructions to complete the data request and data reception of the transmitter; detect whether the data is received or not; detect whether the parsed data is correct; output the data to the display interface.

The collected standard elevation value is obtained by the device, and the algorithm formula of the fast continuous vialogis as shown in the following formulas 2 and 3, and the standard deviation of the evenness is obtained according to the algorithm.

$$d_i = \left(d_{0(i-15)} + d_{0(i+15)}\right)/2 - d_{0i} \tag{2}$$

$$\sigma_i = \sqrt{\frac{\sum_{i=1}^{N}(d_i - \bar{d})^2}{N-1}} \tag{3}$$

Where: σ_i— —the calculation value of the surface evenness of each calculation interval (mm); d_{0i}— —the original value (mm) of the road surface concave-convex deviation displacement acquired at a certain distance with 100m; d_i — —100m as a calculation interval , the correction value of the road surface unevenness deviation (mm) collected at a certain distance; $\bar{d}$— —the average value (mm) of the N road surface concave and convex deviation displacement values collected in a calculation interval; *N*— —a calculation interval is used for calculation The number of test data for standard deviation.

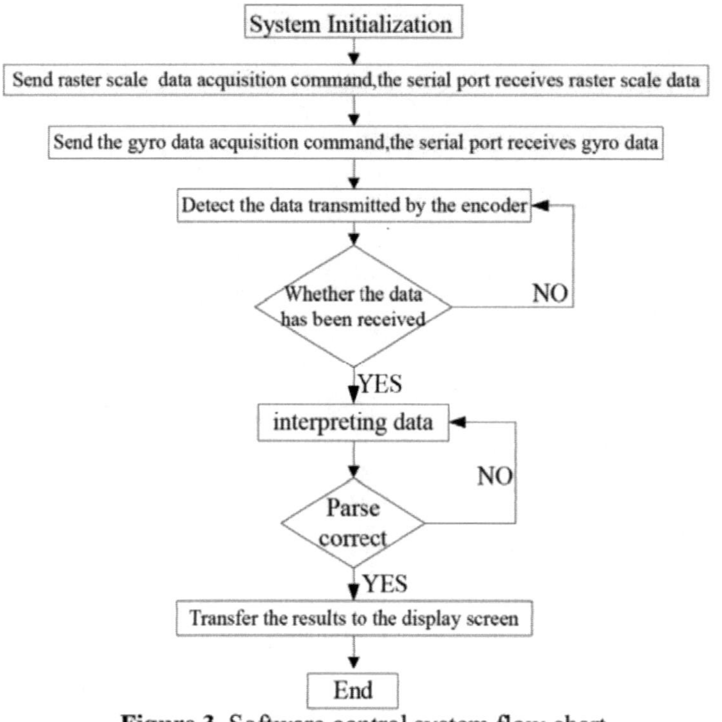

Figure 3. Software control system flow chart.

5. Test results

In accordance with the requirements of JJG (traffic) 024-XXXX continuous vialog (draft for comment), in order to obtain the surface evenness standard value, the two-wheel trace longitudinal elevation data of the test lane must be measured, and the sampling interval is 100mm. The specific plan is as follows:

Use a total station and chalk to draw a 100m long straight line on the standard road, mark the start and end points, mark the points at 100mm intervals along the starting point of the test line. The relative elevation data of each point of the test line is measured by a precision level. During the measurement process, the difference in line of sight between the front view and the rear view (or middle view) should be strictly controlled. The measurement length per station should be 10m when measuring, the level gauge is set in the middle, and the vertical distance of the erection position from the test line is not less than 3.75m. The section is measured with an automatic surface evenness measuring instrument and the elevation value is output. Calculate the surface evenness standard values of the two devices according to the formula (2) and (3). The developed equipment was verified through experiments. The standard values of this equipment were compared with the level equipment, and the results are shown in Table 1. As can be seen from the table, the device error is about 0.01 mm at maximum, which can achieve the intended purpose.

Table 1. Comparison of measurement errors between fast continuous vialog and level.

Standard values	Mcasurcmcnts	Errors	Standard values	Mcasurcmcnts	Errors
1.304706	1.345805	0.041099	1.088615	1.111093	0.022478
1.024592	1.053903	0.029311	1.228306	1.2437	0.015394
1.156418	1.161975	0.005557	1.513746	1.530894	0.017149
2.420903	2.432235	0.011332	1.0246	1.055366	0.030766

6. Conclusion

The fast continuous vialog designed in this paper adopts the contact detection method combining the contact distance measuring device and the gyroscope correction device, and uses the grating scale, the gyroscope and the encoder to perform data sampling, receiving and analysis, and has a small sampling interval. High precision, small error, simple operation, light weight and low price. The fast continuous vialog fundamentally improves the detection efficiency of the continuous vialog and reduces the project acceptance cost. The research of the rapid continuous vialog can provide valuable experience for the research of other vialogs in the transportation industry, and has a positive effect on promoting the inspection of the transportation industry.

Reference

[1] Wang Gang,Yu Shanshan.2017.Transpoworld.36.032:72-73
[2] érgio Pacífico Soncim; Igor Castro Sá de Oliveira; Felipe Brandão Santos; Carlos Augusto de Souza Oliveira;2018. Road Materials and Pavement Design.1448-1457.
[3] Lu Sun. 2003.Mathematics, Computers in Simulation. 2(61):77-88.
[4] Zhang Yi, Ma Ronggui, Ding Hui.2007.Transportation and computer. 25(5):13-16.
[5] Rahma Ktari; Fazia Fouchal; Anne Millien;2017. Christophe Petit; European Journal of Environmental and Civil Engineering.27-42.
[6] Xu Yan.2019.Science & Technology Ecnony Market. 415.1:14-16.
[7] Iván Moyano.2016. Mathematics of Control, Signals, and Systems. Vol.28 (4)

Research on Heat Sissipation System Based on High Power LED-UV Curing Equipment

Li Haichao, Li Jinyao*, Xu Jingxiang, Zhang Mingming, Cai Jifei and Zhang Guishan

School of mechanical and electrical engineering, Beijing Institute of Graphic Communication, Daxing District, Beijing, China

*Email: 346448264@qq.com

Abstract. LED-UV printing curing system is a new green printing curing technology for curing UV ink. Compared with the traditional UV curing device UV-LED curing device, the light source can be stabilized and ready to use without preheating, and the energy consumption is lower. It does not produce ozone and voc, and it is a green curing method. However, since a part of the energy of the LED chip is lost in the form of heat, the temperature of the PN junction is increased to lower the service life and light decay. The heat sink model was established by Solid Works and simulated analysis was performed using Fluent. A heat dissipation system for high-power LED-UV curing device was designed based on the simulation results.

1. Introduction

LED-UV curing is an emerging printing and curing technology. Compared with traditional UV mercury lamps and halogen curing methods, high-power LED-UV has high efficiency, low temperature, convenient installation, stable light source, ready-to-use and energy consumption. Low, green and environmentally friendly, ultra-low noise does not produce many advantages such as ozone and heavy metal pollution. Because of its advantages, LED-UV curing technology has been applied to the fields of screen printing, inkjet printing, flexible printing, lithography and the like.

However, because some of the input energy of the LED chip will be lost in the form of thermal energy, this will cause the temperature of the PN junction of the chip to rise, which will lead to a series of negative effects, such as the failure of the light failure to meet the curing requirements and the greatly reduced service life [1]. So how to improve the heat dissipation efficiency of the system has become one of the urgent problems to be solved in high-power LED-UV curing systems.

2. LED-UV curing system cooling method

At present, the heat dissipation methods widely used in LED-UV curing systems at home and abroad mainly include three types: air-cooling heat dissipation, heat pipe cooling, and liquid cooling. Air-cooling heat dissipation usually uses heat-dissipating materials to make heat-dissipating fins, and heat-dissipating fins and LED chip substrates are bonded together by silicone grease, and air is used as a cooling medium for heat transfer [2]. The performance of air-cooling heat dissipation is closely related to the number of heat dissipation fins, the spacing between the fins, and the shape of the heat dissipation fins. The air-cooling heat dissipation structure is simple, the energy consumption is low, but it is easily interfered by the ambient temperature. Due to the low thermal conductivity of air, air-cooling heat dissipation cannot meet the heat dissipation requirements of high-power LED-UV curing systems. The heat pipe heat dissipation device is composed of a liquid absorbing core, a sealing pipe and a steam passage. When the heat pipe radiator operates, the evaporation passage absorbs heat generated by the heat source, and the heat causes the liquid in the liquid suction core tube to vaporize. The steam with heat moves from the high temperature section of the heat pipe radiator to the low temperature section, and the steam condenses into a liquid when the low temperature section absorbs the heat of the high temperature section steam. The condensed liquid is returned to the high temperature section through the capillary action of the wick on the tube wall, and the heat is continuously dissipated through the above cyclic working mode [3]. This type of heat dissipation is based on purely physical means, without the need for additional external power, low equipment costs, no need for personnel to operate at a time, and high reliability due to metal manufacturing, which requires no maintenance. However, since the erection of the LED-UV curing device needs to be adapted to the compact structure of the printing press, the heat pipe heat dissipating device is usually large in size and cannot meet the requirements. Liquid cooling this heat dissipation method works by circulating a coolant along a cooling pipe that is in close contact with the LED substrate under the push of a cooling pump to remove heat for heat dissipation. The liquid cooling system usually uses water as the cooling liquid. Because the water has a thermal conductivity much larger than that of the air, it can efficiently absorb the heat generated by the LED substrate, and can meet the heat dissipation requirements of the high-power LED-UV system. This heat dissipation method requires external refrigeration equipment to assist the temperature difference between the coolant temperature and the room temperature to be no more than 5 ° C to avoid condensation damage to the LED substrate. The structure of the heat sink is also relatively more complicated, and the overall cost is higher. However, the heat dissipation efficiency is high, the effect is small, the volume is small, and no noise is generated [4]. Therefore, liquid cooling is used as the heat dissipation method of the heat dissipation system.

3. Thermal performance parameters of the heat dissipation system

The heat dissipation system consists of a light source array, copper foil pads, an aluminum nitride substrate, silicon grease, and a heat dissipation plate. Heat is transferred from the PN junction in the light source array to the heat dissipation substrate. A schematic diagram of the components of the heat dissipation system is shown in Figure 1.

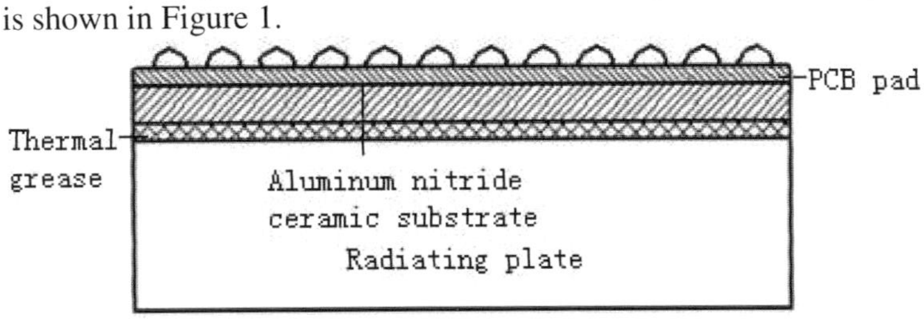

Figure 1. Schematic diagram of the components of the heat dissipation system

The heat in this system will migrate in the form of heat conduction due to the temperature gradient in each component. The thermal resistance values R_1, R_2, R_3 and R_4 respectively represent the light source array, PCB pad, aluminum nitride substrate and heat dissipation paste (silicone grease). Because the packaging material and structure of LED determine the thermal resistance of each lamp bead to be 3.5℃/W, R1=0.00417℃/W is calculated by the light source array in parallel. Since the distance between individual lamp beads in the light source array is extremely small, it is ignored in the calculation, and the substrate area is directly used as the heat conduction area for calculation. The substrate area is S=0.01461 m^2, and the dimensions, thermal resistance and thermal conductivity of other components are shown in Table 1.

Table 1. Thermal parameters of various components in heat conduction

Part name	PCB pad	Aluminum nitride ceramic substrate	Thermal grease
size(mm)	0.08	0.675	0.225
Thermal resistance(℃/W)	0.000015	0.000186	0.003345
Thermal conductivity(W/m• K)	368	247	3.8

The power of each lamp bead in the light source array is about 1.7W, and the total power of the system is 1360W with a total of 800 lamp beads in the light source array. The thermal energy conversion rate of the lamp bead is 30%, and the power of the light source array due to heat generation is P=408W. The total thermal resistance during the entire heat transfer process is:

$$R = R_1 + R_2 + R_3 + R_4 = 0.007716\ (^oC/W) \tag{1}$$

The temperature difference between the heat sink and the light source array:

$$\Delta T = P * R = 408W * 0.007716^oC/W = 3.148^oC \tag{2}$$

4. Heat dissipation structure design

At present, the main way of designing the heat dissipation structure is to establish the model and thermal analysis by using software, and optimize the design through the model of the improved heat dissipation structure [5-7]. The simulation results are analyzed by orthogonal test, and finally the result data is analyzed to complete the optimal design of the heat dissipation structure. In this paper, Solid Works is used to model the heat sink, and then ICEM CFD is used for structural meshing. Finally, the simulation boundary conditions and initial conditions are set by ANSYS Fluent to solve the final analysis conclusion.

The heat sink in this paper adopts the design of baffled parallel cooling channel. The design has only one inlet and one outlet. When the coolant enters from the inlet due to the cooling fins in the cooling channel, the coolant will flow in parallel and flow into each small. Cooling track. The separation of the heat dissipation fins not only increases the contact area between the coolant and the heat sink, but also increases the heat dissipation efficiency. It also increases the turbulence intensity of the coolant in the cooling passage to further enhance the heat transfer effect, thereby making the overall performance of the cooling plate larger. Improvement. The parallel cooling passages have the same size structure, so that the flow rate and the flow rate of the coolant are substantially the same, and the heat dissipation uniformity of the heat dissipation plate is better. The inlet cross-sectional area of the heat sink is 0.785cm2, and the flow rate of the coolant is set to 30m/min, 60m/min, 90m/min, 120m/min, 150m/min, 180m/min, 210m/min, corresponding to the coolant. The flow rates were 2.4 L/min, 4.8 L/min, 7.2 L/min, 9.6 L/min, 12.0 L/min, 14.4 L/min, and 16.8 L/min, respectively. The parameter information and initial conditions used in the software simulation are shown in Table 2. The maximum temperature of the surface of the heat sink at different coolant flow rates is as shown in Table 3. The highest temperature profile of the heat sink surface at different flow rates See Figure 2.

Table 2. Simulation parameters and initial conditions

Coolant initial temperature(K)	Effective heat dissipation area(m^2)	Heat flux(W/m^2)	Cooling hydraulic pressure(Pa)	Lamp beads generate thermal power(W)
300	0.01375	30029.2	102136	408

Table 3. Peak temperature changes of coolant entering the heat sink at different speeds

Coolant flow rate(m/min)	30	60	90	120	150	180	210
Heat sink peak temperature(K)	313.4136	311.6216	310.8845	310.3834	309.9634	309.7658	309.5642

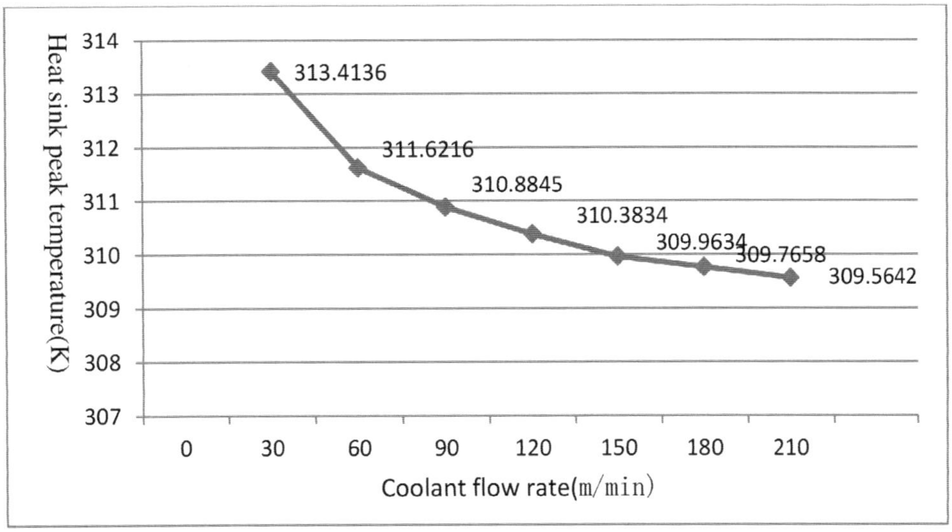

Figure 2. The highest temperature profile of the heat sink surface at different flow rates

Figure 3. Heat sink temperature profile

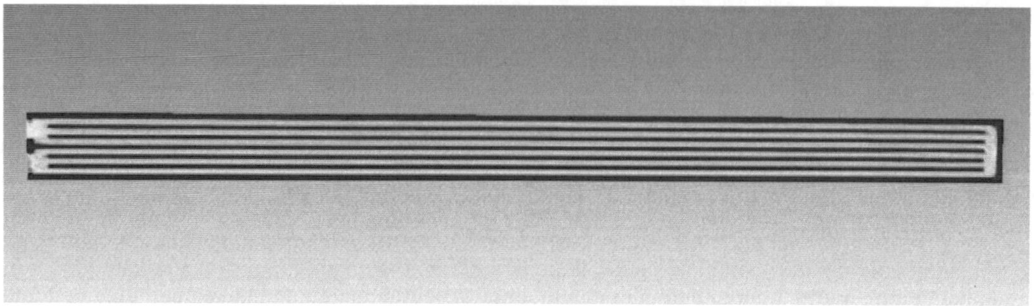

Figure 4. Heat sink cooling channel cross-section velocity distribution

As shown in the temperature profile of the heat sink in Figure 3, the temperature of the heat sink is basically uniform. Only the temperature at the inlet end of the coolant is slightly lower and the temperature at the outlet end of the coolant is slightly higher. The overall temperature of the heat sink is consistent. As shown in the velocity profile of the cooling plate cooling section of Figure. 4, the flow rate of each coolant to be cooled is basically the same, and the flow rate of the coolant is good. According to the highest temperature profile of the surface of the heat sink at different flow rates in Figure 2, the temperature of the surface of the heat sink is inversely proportional to the flow rate of the coolant, but the higher flow rate also means an increase in energy consumption because of the optimal coolant inlet flow rate. It should be selected between 90m/min and 120m/min, and the corresponding flow rate is 7.2L/min~9.6L/min. The maximum temperature of the heat sink is about 310.85K, and the PN junction temperature of the light source array is the sum of the best temperature of the heat sink and the temperature rise of the heat conduction path. The maximum temperature T=(310.88+3.25)K=314.1K, which is about 41.3°C, and the temperature of the PN junction is lower than 50°C. The heat dissipation structure meets the design requirements.

5. Summary

In this paper, a heat dissipating device that can be used in the LED-UV curing system is designed. The air cooling method is adopted as the heat dissipating method by comparing the air cooling heat dissipation, the heat pipe heat dissipation and the liquid cooling. The heat resistance of each part of the heat dissipation system is analyzed, and the heat dissipation system with good heat dissipation performance is designed by modeling and finite element analysis.

Acknowledgments

First of all, I would like to thank the scholars cited in this article for their monographs. Without the inspiration and help from the research results of these scholars, I am unable to complete this article. Secondly, I would like to thank the other authors in this article for their help in my research. Finally, I would like to thank the Collaborative Innovation Center - Green Printing and Packaging Security (Grant No.04190118003/001)and Beijing Municipal Education Commission - Key Technology Research of UV-LED Array Ink Curing Light Source System (Grant No.04190118002/019) and Beijing Education R&D program (Grant No.KM20161005009) project for financial support to enable me to complete the research.

References

[1] Liu Jianxin. Analysis of Thermal Performance of High Power LED Lamps[J].Energy Conservation, 2019(02):31-32.

[2] Ren Hang, Cui Xiaotong, Wang Dadong. *Transient Performance Analysis and Test of UV-LED Air Cooling System*[J]. Energy Engineering, 2016(03): 69-73.

[3] Li Hongchuan, Ji Xianbing, Zheng Xiaohuan, Xu Jinliang. *Parallel multi-channel high-power LED loop heat pipe radiator*[J]. Optoelectronics·Laser, 2015, 26(01): 48-53.

[4] Wang Jie, Li Jian, Fan Bingfeng, Wang Gang. *Design and simulation optimization of high power UV LED water cooling[*J]. Mechanical Design & Manufacturing, 2018(01): 212-214+218.

[5] Pan Yu, Xiang Wenjiang. *Structural design of high-power LED light radiator* [J]. Shandong Industrial Technology, 2018 (05): 7-8.

[6] Zhang Qi, Chen Xu. *Thermal Analysis and Heat Dissipation Structure Design of LED Street Lights*[J]. Electronics and Packaging, 2009, 9(05): 44-48.

[7] Ye Xiangping, Liu Wei, Huang Qingsong. *Simulation and Optimization of High Power LED Heat Dissipation System Based on Flat Heat Pipe*[J]. Electronic Components and Materials, 2017, 36(10): 41-45.

Research Progress on Water-based Ink Drying Technology

Xu Jingxiang*, **Li Jinyao, Li Haichao, Zhang Mingming and Cai Jifei**

School of mechanical and electrical engineering, Beijing Institute of Graphic Communication, Daxing District, Beijing, China

*E-mail:1149843383@qq.com

Abstract. Water-based ink is an environmentally friendly ink that does not contain volatile organic compounds. It conforms to the development tenet of modern society and it doesn't produce solvent volatilization and solvent residue. Therefore, it has great application prospects in the field of packaging printing where relatively strict sanitary conditions are required, however, due to slow drying speed, high energy consumption, low adhesion, etc., there are still insufficient development and application in the market. There are still some shortcomings in development and application. By searching domestic and foreign materials and literature, this paper studied various drying methods of water-based inks, and analyzed the drying efficiency of water-based inks under various drying technologies. The drying efficiency of the aqueous ink is compared under different drying techniques, and the drying methods include hot air drying, infrared drying, microwave drying, and auxiliary drying, By comparing the advantages and disadvantages of different drying methods, it provides a new idea for the efficient and energy-saving of dry water-based inks.

1. Introduction
Plastic flexible packaging printing (intaglio printing typically) produced VOCs emissions from the largest (of the total emissions of VOCs 32.49%)[1].Because water-based inks do not contain volatile organic solvents,and low cost,they have broad application prospects in food packaging, flexible printing, gravure printing[2],China's flexible packaging sector uses only 35% of water-based inks[3],the main limitation of water-based ink application is its drying speed. The solution method has the following two directions. On the one hand, it can increase the drying speed to improve water-based inks additives [4]. On the other hand,it is possible to conduct research from the drying method and to study different drying techniques.Research on different drying technologies to improve the drying speed of water-based inks and reduce energy consumption, so as to achieve efficient and energy-saving green printing.

2. Infrared drying
Infrared drying is a kind of radiant heating. The infrared light emitted by the infrared radiator (light source) is absorbed by the material in the form of molecular (atomic) resonance to achieve the purpose of heating the object. This radiant heating method is a very efficient drying method by directly matching the surface of the object and a certain depth with a matching wavelength, selective penetration capability, and orientation. The spectrum of the high-life medium wave radiator is very well matched to the absorption line of water, so it is highly efficient and economical and durable in aqueous ink drying .

Selim [5] proposed to use infrared drying technology to dry in order to solve the problem of ink penetration into paper and evaporation in the course of the drying process of water-based inks.At the same time, a set of drying models was established.It found that for plain paper used in the office environment, the rate of penetration was at least 20x higher than the rate of evaporation.Thus,Selim believes that the rate of ink drying can be predicted by the paper's penetration curve.

Seaton [6] in order to improve the drying efficiency of water-based inks, accelerate the use of water-based inks in the field of flexible printing, the infrared drying technology is cited in the flexible printing process, the experimental results show that the use of infrared drying in flexible printing, not only can improve the drying effect, At the same time, in the case of security conditions, you can also produce papers with improved dot gain.

The biggest difference between the traditional ink and the water-based ink is the difference in the binder. The dye is fixed on the paper by the binder. The Connecting material of water-based inks is basically made of water, and the drying process is simplified. Shi Xiangdong [7] proposed that for infrared drying equipment, the biggest disadvantage of infrared drying equipment is that the installation of the device is complicated and the investment cost is too high.

Imakoma [8] used the idea of temperature change to distinguish the previous measurement of moisture loss from the previous use and to change the temperature change of the substrate during the drying process. The experimental results show that the temperature change during the drying process can better fit the drying rate curve of the infrared water-based paint.

When Manabu [9] studied inkjet printing technology, it was found that printing speeds exceeding 200 m/min caused wrinkles in the non-printing area of the paper, which affected the drying effect. The experimental results show that the traditional infrared drying technology is applied to inkjet printing, which has no effect on the classical printing effect. At the same time, this method can reduce the breakage rate of paper, and is more efficient and economical.

The HP L65500 printer [10] used a small infrared device to dry HP's own water-based latex ink. The classic performance of the L65500 model printer maintains a drying time of 30 s at a drying temperature of 90 °C, allowing the molecules to diffuse enough time to produce a latex film.

3. Hot air drying

Compared with the high cost of infrared drying, the hot air drying equipment is simple to install and easy to maintain, and has become the mainstream drying processing equipment in the market. In the hot air drying process, the transfer of thermal energy, the evaporation of liquids such as water and solvents, and the barrier of solid materials to fluid evaporation will affect the drying speed of the ink. As the hot air of the drying body, its speed and temperature are the key factors determining the drying efficiency. The higher the speed, the higher the temperature and the higher the drying efficiency, but the corresponding fan power and heating power will also increase. Therefore, the designer of the drying equipment needs to understand the basic drying process to determine the reasonable nozzle size, fan power and hot air temperature.

Imakoma [11] applied water-based paint on glass and passed 10m/s hot air on both sides of the glass to judge the water loss rate of the product before and after drying by the temperature change in the experiment, and then judge the drying effect. Experimental results show that, at a constant speed of hot air, in the middle of the hot air drying of the product to ensure heat uniformity.

Cao Le [12] analyzed the parameters of gravure hot air drying. At the initial dose of 50%, The printing speed of 120m /min, width 1000mm, the drying zone is in the field, speed of the fan under 1100r /min, and for different print convection hot air temperature drying experiment to find the factors affecting the drying of the print, to establish a multi-form A mathematical model of regression analysis to optimize parameters. The experimental data shows that the faster the hot air speed, the better the printing effect, and doesn’t affect the printing quality.

Based on nonlinear regression, Jia Chunjiang [13] used several typical drying models to simulate in the case of hot air temperature of 50 ° and hot air velocity of 6 m/s. The results show that on the Logarithmic model, a new set of page models is obtained after modification. The experimental results are in good agreement with the model predictions, and have a good reference effect on the drying rate and effect prediction of water-based inks.

Plavnik [14] used acoustic vibration and hot air drying technology to dry water-based inks and other water-based materials. The results show that when drying water-based inks, lower heat is allowed at higher paper web speeds. Air (lifting efficiency) operating temperature performs this process. It can be used in narrow and wide width applications, expanding the range of applications for waterborne inks.

In order to improve the drying speed of the aqueous ink, Nowaczek [15] combined hot air drying with sonic vibration. The experimental results show that the sonic vibration can accelerate the molecular motion of the binder in the early stage of the aqueous ink in the early stage of the drying process. Reduced hot air energy consumption and improved drying efficiency.

4. Microwave drying

Microwave drying is also a way to dry the ink by heating. Under the action of the microwave electromagnetic field, the heated medium molecules become polarized molecules with positive and negative charges. Because microwaves are non-ionizing, unlike ionizing radiation, they can interact with dielectric materials to generate heat by agitating molecules in an alternating electromagnetic field. Microwave energy (in the heat application) by said hollow tube is a metal waveguide transmission, to avoid excessive radiation. They are reflected by metal objects, absorbed by some dielectric materials, and transmitted without significant absorption by other dielectric materials. For example, water, carbon and foods with high water content are good microwave absorbers; glass, ceramics And most thermoplastic materials allow microwaves to pass through with little or no absorption. This is the mechanism by which the water-based ink is dried.

Marios [16] compared several drying methods common to water-based inks, hot air drying, the IR drying, microwave drying rate and energy consumption, finishing the results shown in Table 1.This experiment proves that microwave drying has obvious advantages over both in terms of drying rate and energy consumption.

Table 1. Comparison of drying time and energy consumption of different methods of drying aqueous inksextend it slightly outside the width of the table. Justify the text. Leave 6 pt of space between the caption and the top of the table.

Ink type	Solvent	Drying method	Drying time	Energy consumption(kWh)
Water-based inks	water	IR drying	30sec	0.0083
		Hot air drying	60sec	0.0238
		Microwave drying	20sec	0.0067

Jaiswar [17] designed a novel microwave absorbing device that can evenly distribute the microwave field strength during the drying process of the aqueous ink and greatly increase the drying speed. The experimental results show that in the 5-50 GHz frequency range , the reflection coefficient bandwidth is less than 15 dB at a bandwidth of 31 GHz on a flexible PC board of 5 mm , that is, the drying effect is best at this time.

Norte [18] designed a new dry structure to speed up the drying speed in order to solve the problem that the time of microwave drying water-based ink is too short to match the high drying speed. Experiments have shown that, depending on its microwave field strength and dry length, the device can dry paper as wide as 60 cm.

KUST [19] has developed a set of microwave drying equipment that can be used in industrial applications. The device can be used for fast drying aqueous inks as well as water based coatings. The study shows that the microwave drying technology can achieve the same speed as the drying instruments commonly used on the market in the case of drying the aqueous ink. At the same time, according to the study by Marios, the energy consumption of microwave drying is much smaller than that of hot air drying.

Trembley [20] proposed the possibility of applying an electromagnetic field to dry aqueous inks. A stable electromagnetic field can produce a dielectric thermal effect. The experimental results show that the electromagnetic field heating can effectively dry the aqueous ink in the case of the gravure printing machine motor 1500r/min. Although this speed is too slow, it can be combined with high-speed hot air drying, infrared drying, and other conventional drying methods to increase the drying speed.

Giles [21] modified the drying facility of the gravure press to address VOC emissions during the printing process. Giles used a paper length of 80 mm and a width of 40 mm to study the relationship between drying time and drying efficiency under constant microwave field strength. The research results show that in the gravure printing machine, the printing speed reaches 60m/s, the energy consumption is 2kw, and the drying time is only 0.0038 seconds, without affecting the printing quality. The study also showed that microwave drying does not cause damage to the non-printing area of the paper.

5. Other drying method

Wei [22] uses a chemical agent to assist the aqueous ink drying process, and develops a temperature-controlled microcapsule desiccant that releases calcium oxide which reacts with water and generates heat when heated in a drying oven of a gravure printing machine, thereby promoting ink drying. The experimental results show that, for the conventional gravure printing machine, the aqueous ink containing no desiccant does not have a good printing effect at a printing speed exceeding 60 m/min, and is not suitable for industrial production, and the aqueous ink to which the desiccant is added exceeds In the case of a printing speed of 80 m/min, a good printing effect is maintained.

The core process of water-based ink drying is that water evaporates from the surface of the ink film. The main factors affecting the drying speed of the aqueous ink are the surface tension of water and the heat of evaporation. Zhang [23] improves the release of water by adjusting the tension of the surface of the system. Increasing the contact area between the ink film and the substrate to achieve the purpose of increasing the drying speed. The results show that the surfactant has a great influence on the drying speed of the aqueous ink. The proper addition of dispersant and wetting agent can promote the drying speed; the addition of the inappropriate amount not only reduces the drying speed, but also the water-based ink. Other properties such as viscosity, fineness, and adhesion fastness have an adverse effect.

In the process of printing on paper, the water-based ink has uneven density, and there are gaps of millimeters to millimeters. The ink drying speed is studied, and these gaps are not negligible. Swerin [24] also studied the effect of surfactants on the drying speed of aqueous inks. The researchers studied several aqueous inks and desiccants commonly found on the market and explored the reaction of cationic fixatives and polyvinyl alcohol binders with pure water binders in inks. The results show that the reaction between cations and polar molecules can increase the drying speed of aqueous inks, but the amount must be appropriate, otherwise the drying speed will decrease, and the adhesion fastness will be greatly reduced.

6. Conclusion

The microwave system effectively and quickly and efficiently dries the water-based inks on the paper. It also compares the environmental benefits of using a microwave-assisted drying system compared to a conventional drying system. The environmentally friendly effect is significant because the microwave drying system offers the possibility of using water-based inks instead of solvent-based inks.

Drying using microwaves can provide significant energy and cost efficiencies, and the quality of rapid offset printing will also increase while using a microwave system for paper drying between prints.

At the same time, the industrialization of microwave equipment and the production restrictions of water-based inks have been solved, which provide a powerful birth condition for microwave drying aqueous ink systems.

Acknowledgments
Here, I would like to thank my teacher, Professor Li Jinyao, for his selfless guidance, and also to Professor Cai Jifei and Zhang Mingming for answering my questions. Thanks to the fund support from the project Collaborative Innovation Center - Green Printing and Packaging Security (04190118003/001) and the project Beijing Municipal Education Commission - UV-LED array ink curing light source system key technology research (04190118002/019) and Beijing Municipal Education Commission General Project(KM20161005009).

References
[1] WANG Xiaofang, LI Zhao, SUN Jianming. Analysis of the current status and application of environmentally friendly water-based inks [J]. Science Technology and Engineering, 2017(15):172-180.
[2] Xin Xiulan , Wei Yana.Research Progress in Water-based Inks [J].*China Printing and Packaging Research*, 2011, 03(3):1-8.
[3] Lai Ho. *The aqueous ink and the development status analysis* [J]. Paper and Paper, 2017 (5).
[4] Sun vibrator, Weixian Fu, Huangbei Qing, et al *Drying factors aqueous plastic gravure ink* [J] Effects Packaging Engineering, 2010 (17): 118-120.
[5] Selim MS, Yesavage VF, Chebbi R, et al. *Drying of Water-Based Inks on Plain Paper* [J]. Journal of Imaging Science & Tech nology, 1997, 41(41): 152-158.
[6] Seaton, Wayne. *Improved line and process graphics* [J]. International Paper Board Industry, 2000 , 43 (9) .
[7] Shi Xiangdong, Cheng *Now often physical drying process and mechanism of a printing ink* [J]. Packaging engineering, 2004, 25 (2).
[8] Imakoma, Hironobu; Takenaka, Kei; Kawano, et al . *Measurement of infrared dryin g rate of coating by an extended temperature-change method* [J]. Kagaku Kogaku Ronbunshu, 2014 , 40 (1) .
[9] Numata M , Sakamoto A , Ogasawara Y , et al. Drying Technology Using Laser Exposure for High-speed Inkjet Printing[C]// NIP & Digital Fabrication Conference. 2013.
[10] Maza J G. HP Designjet L65500 Drying and Curing Systems[C]// Nip & Digital Fabrication Conference. 2012.
[11] Imakoma H, Nagaoka S, Takigawa T. *Measurement of drying rate of water-based coat by temperature change method* [J]. Kagaku Kogaku Ronbunshu, 2007, 33(6): 586-592.
[12] Cao Yue, Wang Anita *Analysis of MATLAB-based gravure hot air drying parameters* [J] Packaging Engineering, 2008, 29 (6): 69-70.
[13] Jia Chunjiang, Chan school, Li Xiao-week *Study based ink and hot air drying mechanism mathematical model* [J] China Printing and Packaging Study, 2011 (3): 47-50.
[14] Plavnik, Gene. *Drying at the speed of sound*[J]. Coating International, 2012,45(2).
[15] Nowaczek E J. The Application of Air-Blowing Ultrasonics in Drying of Water-based Inks on Films and Foils[J]. American inkmaker, 1997, 75(9).]
[16] Marios T, Eugenia P, Anastasios P, et al. *Using microwave drying systems in the Graphic Arts. Modern solutions for environmental industrial applications* [J].
[17] Jaiswar R, Mederos-Henry F, Dupont V, et al. *A thin ultra-wideband microwave absorbing structure printed on flexible substrate with resistive-ink made of multiwall carbon-nanotube*[C]//2017 11th International Congress on Engineered Materials Platforms for Novel Wave Phenomena (Metamaterials). IEEE, 2017: 160-162.
[18] Norte D. *Human Exposure Design Considerations For Machine Directed Microw ave Dryers For Drying Ink On Paper Webs For The Inkjet Printing Industry* [C]//2018 IEEE Symposium on Electromagnetic Compatibility, Signal Integrity and Power Integrity (EMC, SI & PI). IEEE ,

2018: 193-198.

[19] Shaohua J , Singh P , Jinhui P , et al. *Recent deve lopments in the application of microwave energy in process metallurgy at KUST*[J]. Mineral Processing and Extractive Metallurgy Review, 2017:1-10.

[20] Trembley J F, LORING C M. Drying printing inks and coatings by radio frequency heating [J]. Tappi, 1969, 52(10): 1847-&.

[21] Saad A A E R E. Environmental pollution reduction by using VOC-free water-based gravure inks and drying them with a new drying system based on dielectric heating[D]. Universität Wuppertal, Fakultät für Elektrotechnik, Informationstechnik und Medientechnik» Medientechnik» Dissertationen, 2008.

[22] Xu Yingjie, Zhang Fuzhong, Wei Xianfu, et al. Study on the efficacy of temperature-controlled microcapsule desiccant for water-based gravure inks[J]. Journal of Beijing Institute of Graphic Communication, 2013(6): 31-33.

[23] Zhang Song, Yang Xijiang, Yang Wei. Effect of Surfactants on Drying Rate of Water-based Inks[J]. Liaoning Chemical Industry, 2010, 39(3): 245-247.

[24] Swerin A, König A, Brandner B, et al. The use of silica pigments in coated media for inkjet printing: Effects of absorption and porosity on printing performance and depth-profiling using confocal Raman spectroscopy[C]//TAPPI Advanced coating fundamentals symposium proceedings. 2008: 178-203.

IOP Conf. Series: Materials Science and Engineering **565** (2019) 012018 doi:10.1088/1757-899X/565/1/012018

Femtosecond Laser Double Pulses Nanofabrication on Silicon

Jin Zhang[1*], Sumei Wang[1], Mengmeng Wang[1] and Zhuyuan Chu[1]

[1] Laser Micro/Nano Fabrication Laboratory, School of Mechanical Engineering, Beijing Institute of Technology, Beijing 100081, P.R. China

[*]Email: zj_1995@126.com

Abstract. The temporal shaping femtosecond (fs) laser is realized by dividing a fs laser pulse into two identical sub-pulses. The double-pulse fs laser can be used to effectively control the initial free electron state such as electron temperature, capacity and density, etc. to improve the surface morphology quality. In this experiment, silicon was fabricated using the double-pulse fs laser. It was found that the average diameter of ablated micro- or nanoholes decreases with the increasing pulse delay (up to 1.5 ps) under the same laser fluence and the phenomenon was explained quantitatively by the plasma model. Furthermore, nanoholes were achieved using the double-pulse fs laser, which cannot be obtained by single pulse laser, and the processing size can be reduced to below 200 nm.

1. Introduction

Depending on unique processing advantages compared to other special processing techniques such as electron beam machining [1], focused ion beam machining [2], and photolithography [3], etc. femtosecond (fs) laser beam machining is an irreplaceable technique in the micro- and nanofabrication field. Due to ultrahigh peak power and ultrashort pulse duration, fs laser processing can be regarded as a 'cold' processing. Meanwhile, relying on its non-linear effect [4], fs laser can process any difficult-to-machine material including metal, dielectric and semiconductor, etc. In addition, it is simple to operate, low-cost and can realize three-dimension processing. Because of ultrashort pulse duration of fs laser, only the free electron states such as free electron capacity, temperature and density, etc. can be affected during the laser irradiation [5]. The double-pulse fs laser can control these properties of free electron by adjusting the laser pulse delay [6]. Furthermore, the subsequent physical processes can be affected and the processing quality can be optimized.

Fs laser temporal shaping technique can be applied in various fields of laser processing. In terms of laser modification [7], the absorption rate of fs laser can be improved using the double-pulse fs laser under the same laser fluence. Then the increased absorption rate leads to the formation of more 3-membered ring structures inside fused silica. Based on more 3-membered ring structures, laser-irradiated fused silica is easier to be etched by HF acid and the etching efficiency is greatly improved. In the laser ablation field, the fs laser pulse sequence can be achieved through pulse shaper, and the ionization process of material is changed by the fs laser pulse sequence. Using the fs laser pulse sequence, a high depth-diameter ratio microhole can be processed on the sapphire substrate, which cannot be achieved by an ordinary Gaussian-like single pulse [8].

In this paper, the temporal shaping [9] was realized by dividing a fs Gaussian-like single pulse into two identical sub-pulses with the same fluence. As a semiconductor, silicon is widely applied in the electronic information field. The double-pulse fs laser was used to process the crystal <111> silicon in this experiment. By changing the pulse delay to control laser-material interaction process, the surface morphology of silicon can be optimized. Compared to the single pulse, it was found that the average diameter of microholes processed by the double pulses is smaller than the single pulse and the diameter became smaller with the increasing pulse delay under the same laser fluence. Using the critical free electron density as the laser ablation criterion [10], the change of laser ablation threshold was predicted by the plasma model [5]. Furthermore, we found that double pulses can achieve more precise machining which cannot be completed under the single pulse, and the diameter of holes can be reduced to below 200 nm.

2. Experimental section

A commercial Spectra Physics Spitfire laser with a central wavelength of 800 nm and a pulse delay of 35 fs was used to ablate the crystal <111> silicon. A single pulse fs laser was divided into two identical sub-pulses with an energy ratio of 1:1 through the beam splitter. The optical path difference of two sub-pulses was adjusted to control the time delay of reaching the silicon surface through the electronic translation stage. The temporal shaping was achieved using the double-pulse fs laser. In this experiment, a single shot pulse was used to ablate these holes on a silicon substrate. The laser was focused by a 10X objective lens. Further, in order to make the hole diameter as small as possible, a 50X objective lens was used to generate the tightly focused fs laser pulse to explore the minimum size. The processing results were observed by cold field scanning electron microscopy (SEM) for detailed data analysis.

3. Results & Discussion

3.1. Double Pulses Nanofabrication

Firstly, a fs laser single pulse was used to ablate the microhole structure on silicon. In order to compare the difference between the double pulses and single pulse, the pulse delay was changed from 0 fs to 2000 fs under the same laser fluence. As shown in Figure 1, microholes were respectively fabricated at the laser fluences of 0.52 J/cm^2 and 0.35 J/cm^2 to ensure the good repeatability. Figure 1a and Figure 1b show the change of surface morphologies at the laser fluence of 0.52 J/cm^2, and the black area in the center of microholes means that the microholes were ablated. However, as shown in Figure 1c and Figure 1d, the white area in the center of microholes means the formation of bump. The laser irradiated area was modified and the laser fluence of 0.35 J/cm^2 was under the ablation threshold. The formation mechanism of bump was attributed to the fact that the liquid density (2.52 g/cm^3) of silicon is larger than the solid density (2.32 g/cm^3).

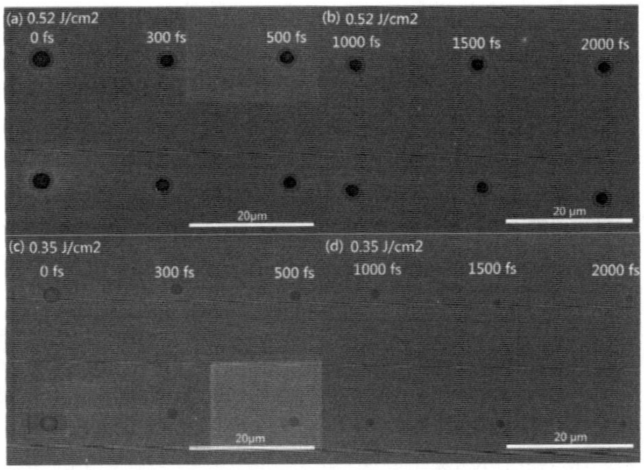

Figure 1. SEM images of silicon surfaces exposed by (a-b) a fs pulse at a fluence of 0.52 J/cm^2 with pulse delays of 0, 300, 500, 1000, 1500 and 2000 fs, respectively, and (c-d) a fs pulse at a fluence of 0.35 J/cm^2 with pulse delays of 0, 300, 500, 1000, 1500 and 2000 fs, respectively. The experimental parameters of holes in each column are uniform.

By counting the diameters of microholes, it was found that the microhole became smaller with the increasing pulse delay and the diameter gradually stabilized as the pulse delay is up to about 1500 fs. Figure 2 shows the trend change of hole diameter as a function of pulse delay.

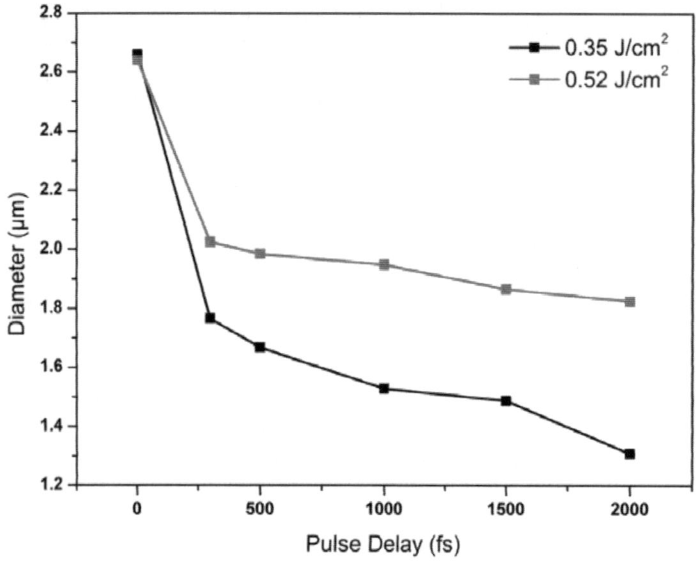

Figure 2. Average diameter of microholes as a function of pulse delay at the double-pulse threshold fluence of 0.52 J/cm^2 and 0.35 J/cm^2, respectively.

3.2. Plasma Model Theory

This phenomenon that the hole diameter decreased with the increasing pulse delay under the same laser fluence is also found in the fs laser ablation of fused silica [6]. Compared to the insulator, the properties of semiconductor silicon are similar to fused silica during the process of laser interaction. Therefore, the plasma model can be utilized to describe the laser-silicon interaction. The direct cause of this phenomenon is that the increase of pulse delay lead to the increase of the ablation threshold of silicon. Due to higher ablation threshold of large pulse delay, the diameter of microhole is smaller than the single pulse under the same laser fluence. Further, depending on the critical electron density ablation criterion, the ablation threshold of silicon can be calculated by the plasma model. When the fs laser double pulse irradiates on the silicon surface, the first sub-pulse excites a large amount of free electrons from the valence band of silicon by photoionization, and the free electron density cannot remain constant during the time interval between two sub-pulses. During the time interval, the recombination of free electron occurs due to direct radiation recombination, Auger recombination, and diffusion of free electron [11]. The free electron density decreases through the recombination process. During the second sub-pulse irradiation, these free electrons act as seed electrons to excite more free electrons through impact ionization, and large amounts of free electrons also are excited through photoionization. Therefore, as the pulse delay increases (up to 1500 fs), the free electrons ionized by the first sub-pulse will recombine more so that the total free electron density naturally become less. According to the critical free electron density ablation criterion, the ablation threshold will become larger.

Furthermore, the plasma model was used to simulate the relationship between the ablation threshold of silicon and pulse delay of double pulses quantitatively. Since silicon is a semiconductor, the valence band is only 1.12 ev [12], which is lower than the 1.5 ev of a single photon energy. This means that the ionization is mainly generated by single photon and two-photon absorption [13], and impact ionization can be ignored under the low laser fluence [14]. Same with fused silica, the recombination term of free electron density can be simplified by $-\frac{n_e}{\tau}$ [15]. The typical equations for calculating the free electron density after the laser irradiation on silicon are given by:

$$\frac{\partial I}{\partial z} = -\alpha_0 I - \beta I^2 - \alpha_{fca} I \quad (1)$$

$$\frac{\partial n_e}{\partial t} = \frac{\alpha_0 I}{\hbar\omega_L} + \frac{\beta I^2}{2\hbar\omega_L} - \frac{n_e}{\tau} \quad (2)$$

where I means the initial optical intensity, z is the spatial coordinate perpendicular to the silicon surface, α_0 and β are single photon and two-photon absorption coefficient, α_{fca} refers to the absorption rate of free electron, n_e is the free electron density and τ describes the relaxation time of free electron, which is about 200 fs for silicon [15].

Different pulse delay means different optical intensity distribution in the time scale. Through the iteration of equation (1) and equation (2), we can calculate the relationship between pulse delay and ablation threshold, which fits well with the experimental data, as shown in Figure 3. The ablation threshold of silicon becomes larger and finally stabilizes with the increasing pulse delay from 0 fs to 2000 fs. Therefore, the average diameter of holes naturally becomes smaller.

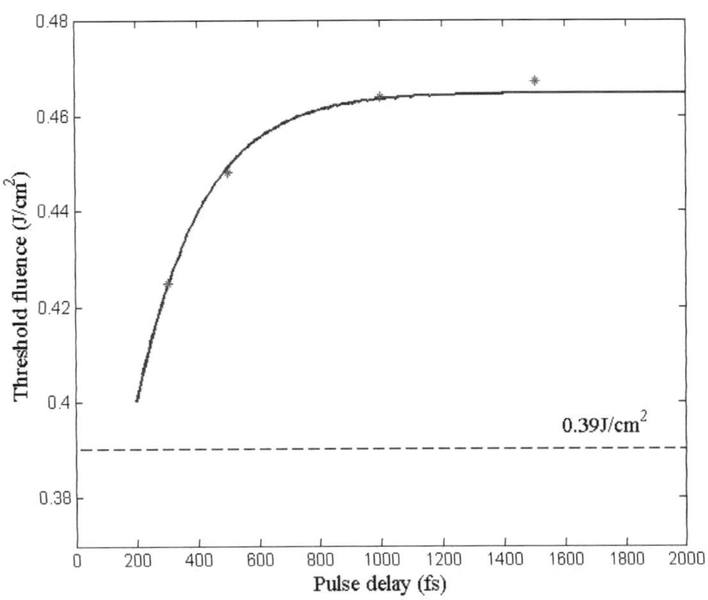

Figure 3. The relationship between ablation threshold and pulse delay is stimulated through the plasma model, and the dot line means the ablation threshold under the single pulse. The asterisks represent the ablation thresholds under different pulse delay.

3.3. Exploration of Minimum Size

In order to ablate enough small hole on silicon using fs laser, a 50X objective lens was used to focus tightly fs laser to obtain waist diameter of 1.57 μm, which is calculated using a method proposed by Liu [16]. By continuously reducing the laser fluence, a hole with diameter 400 nm was processed at laser fluence of 0.52 J/cm^2 under the single pulse, which is larger than the ablation threshold fluence under single pulse, as shown in Figure 4. The smallest hole was processed under single pulse and silicon cannot be ablated under double pulses at the same laser fluence, which proved the ablation threshold under double pulses was larger than that under single pulse. Using the double pulse, it can be found that the diameter of nanohole was about 325 nm at the laser fluence of 0.62 J/cm^2 and pulse delay of 2000 fs (Figure 5a). A nanohole with diameter of about 200 nm was processed at the laser fluence of 0.57 J/cm^2 and pulse delay of 300 fs, as shown in Figure 5b, which is the smallest hole at present and cannot be achieved under the single pulse.

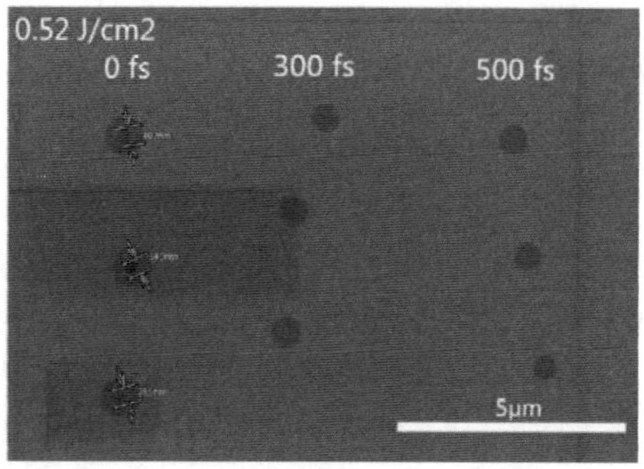

Figure 4. SEM images of silicon surfaces exposed by a fs pulse at a fluence of 0.52 J/cm^2 with pulse delays of 0, 300, and 500 fs, respectively. The experimental parameters of holes in each column are uniform.

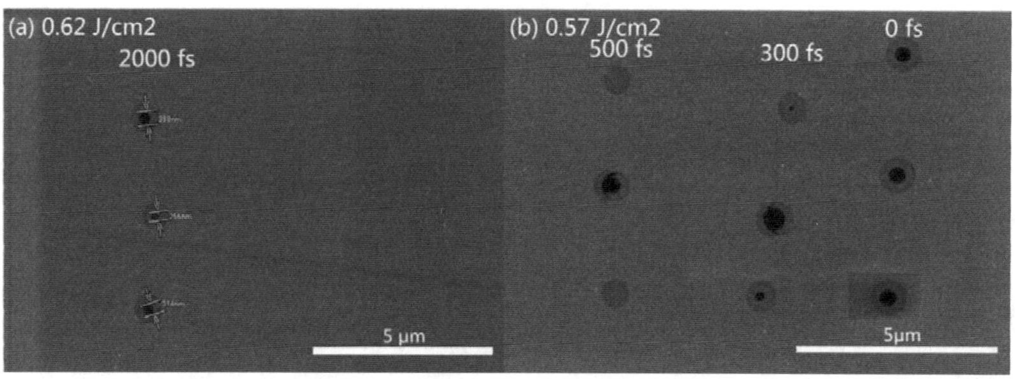

Figure 5. The minimal size of hole fabricated at the laser fluences of 0.62 J/cm^2 (a) and 0.57 J/cm^2 (b), respectively. The experimental parameters of holes in each column are uniform.

4. Conclusion

In conclusion, we use the double-pulse fs laser to achieve the temporal shaping of fs laser. Compared with the traditional single pulse, double pulses can precisely control the properties of free electron such as temperature, capacity and density, etc. by adjusting the pulse delay. By ablating microhole structure on silicon material using double-pulse fs laser, it was found that the average diameter of holes decreases with the increasing pulse delay from 0 fs to 2000 fs under the same laser fluence. Based on the plasma model, this phenomenon can be explained quantitatively from the prospective of ablation threshold. Furthermore, the processing accuracy can be improved using the double-pulse fs laser.

Acknowledgments

The authors gratefully acknowledge the financial support of the National Key R&D Program of China (Grant No. 2017YFB1104300) and the National Natural Science Foundation of China (NSFC) (Grant No. 51575053).

IOP Conf. Series: Materials Science and Engineering **565** (2019) 012018 doi:10.1088/1757-899X/565/1/012018

References

[1] Corbierre M K, Beerens J and Lennox R B 2005 *Chem. Mater.* **17** 5774–9
[2] Melngailis J, Mondelli A A, Berry I L and Mohondro R 1998 *J. Vac. Sci. Technol. B* **16** 927–57
[3] Chou S Y, Krauss P R and Renstrom P J 1996 *J. Vac. Sci. Technol. B* **14** 4129–33
[4] Levis RJ, Menkir GM and Rabitz H 2001 *Science* **292** 709–713
[5] Jiang L and Tsai H L 2005 *Int. J. Heat Mass Transfer* **48** 487–499
[6] Jiang L and Tsai H L 2005 *Appl. Phys. Lett.* **87** 151104
[7] Zhao M J, Hu J, Jiang L, Zhang K H, Liu P J et al. 2015 *Sci. Rep.* **5** 13202
[8] Javier H, Nadine G, Jan S, Michelina S, Bastian Z, et al. 2015 ACS *Appl. Mater. Interfaces* **7** 6613–6619
[9] Lan Jiang, An-Dong Wang, Bo Li, Tian-Hong Cui and Yong-Feng Lu. 2018 *Light: Science & Applications* **7** 17134
[10] Stuart B C, Feit M D, Rubenchik A M, Shore B W and Perry M D 1995 *Phys. Rev. Lett.* **74** 2248–2251
[11] Sundram K S and Mazur E 2002 *Nature Mater.* **1** 217–224
[12] Bristow D A, Rotenberg N and Driel M H 2007 *Appl. Phys. Lett.* **90** 191104
[13] Sokolowski T K and Von der Linde D 2000 *Phys. Rev. B* **61** 2643–2650
[14] Deng Y P, Xie X H, Xiong H, Leng Y X, Cheng C F, Lu H H, Li R X and Xu Z Z 2005 *Opt. Express* **13** 3096–3103
[15] Li M, Menon S, Nibarger P J and Gibson G N 1999 *Phys. Rev. Lett.* **82**, 2394–2397
[16] Liu J M 1982 *Opt. Lett.* **7** 196-198